Agro-Industries: Paper, Pulp, Rubber and Tea

Agro-Industries: Paper, Pulp, Rubber and Tea

Prof. Sriniwas Puri

RANDOM

RANDOM PUBLICATIONS

NEW DELHI (INDIA)

Agro-Industries: Paper, Pulp, Rubber and Tea

ISBN 978-93-5111-768-1

Published in 2016 in India by

RANDOM PUBLICATIONS

4376-A/4B, Gali Murari Lal, Ansari Road
New Delhi-110 002
Phone : +9111-43580356, 011-23289044, 011-43142548
e-mail: sales@randompublications.com,
info@randompublications.com, randomexports@gmail.com

Type Setting by : Friends Media, Delhi-110089
Printed at : Sanat Printers

Preface

The agro-industry mainly comprises of the post-harvest activities of processing and preserving agricultural products for intermediate or final consumption. It is a well-recognized fact across the world, particularly in the context of industrial development, that the importance of agro-industries is relative to agriculture increases as economies develop. It should be emphasized that 'food' is not just produce. Food also encompasses a wide variety of processed products. It is in this sense that the agro-industry is an important and vital part of the manufacturing sector in developing countries and the means for building industrial capacities.

The Indian Paper Industry accounts for about 2.6% of the world's production of paper. The estimated turnover of the industry is INR 50,000 crore approximately and its contribution to the exchequer is around INR 4500 crore. The industry provides employment to more than 0.5 million people directly and 1.5 million people indirectly. Most of the paper mills are in existence for a long time and hence present technologies fall in a wide spectrum ranging from oldest to the most modern.

Pulp is a lignocellulosic fibrous material prepared by chemically or mechanically separating cellulose fibres from wood, fiber crops or waste paper. Although the development of tea as a thriving industry in India has been more recent, historical records indicate the prevalence of tea drinking in India since 750 BC. In the 16th century, it was also observed that people in India were preparing a vegetable dish using tea leaves with garlic and oil.

The ultimate thrust towards commercial tea cultivation was driven by the incorrigible tea addiction of the British. Due to this addiction, they were buying it in large quantities from the Chinese. By 1750, they were purchasing millions of pounds of tea every year from China. Even though they managed to counterbalance it with the opium trade to some extent, it was obvious that their tea addiction was getting exorbitantly expensive and unsustainable.

The book will suit to the needs of students, teachers, scholars and general readers.

– Author

Contents

1

Agro Industry Scenario

AN INTRODUCTION

The agro industry is regarded as an extended arm of agriculture. The development of the agro industry can help stabilise and make agriculture more lucrative and create employment opportunities both at the production and marketing stages. The broad-based development of the agro-products industry will improve both the social and physical infrastructure of India. Since it would cause diversification and commercialization of agriculture, it will thus enhance the incomes of farmers and create food surpluses.

The agro-industry mainly comprises of the post-harvest activities of processing and preserving agricultural products for intermediate or final consumption. It is a well-recognized fact across the world, particularly in the context of industrial development, that the importance of agro-industries is relative to agriculture increases as economies develop. It should be emphasized that 'food' is not just produce. Food also encompasses a wide variety of processed products. It is in this sense that the agro-industry is an important and vital part of the manufacturing sector in developing countries and the means for building industrial capacities.

The agro Industry is broadly categorised in the following types:

(i) Village Industries owned and run by rural households with very little capital investment and a high level of manual labour; products include pickles, papad, etc.

(ii) Small scale industry characterized by medium investment and semi-automation; products include edible oil, rice mills, etc.

(iii) Large scale industry involving large investment and a high level of automation; products include sugar, jute, cotton mills, etc.

The development of agro-based industries commenced during pre-independence days. Cotton mills, sugar mills, jute mills were fostered in the corporate sector. During the post-Independence days, with a view to rendering more employment and using local resources, small scale and village industries were favored.

The increasing environmental concerns will give further stimulus to agro based industries. Jute and cotton bags, which have begun to be replaced by plastic bags, have made a comeback. It is the right time to engage in mass production of low cost jute/cotton bags to replace plastic bags.

The agro industry helps in processing agricultural products such as field crops, tree crops, livestock and fisheries and converting them to edible and other usable forms. The private sector is yet to actualize the full potential of the agro industry.

The global market is mammoth for sugar, coffee, tea and processed foods such as sauce, jelly, honey, etc. The market for processed meat, spices and fruits is equally gigantic. Only with mass production coupled with modern technology and intensive marketing can the domestic market as well as the export market be exploited to the fullest extent. It is therefore imperative that food manufacturers understand changing consumer preferences, technology,With modernization, innovation and incorporation of latest trends and technology in the entire food chain as well as agro-production, the total production capacity of agro products in India and the world is likely to double by the next decade.

India is the second largest producer of food in the world. Whether it is canned food, processed food, food grains, dairy products, frozen food, fish, meat, poultry, the Indian agro industry has a huge potential, the significance and growth of which will never cease.

Sea fishing, aqua culture, milk and milk products, meat and poultry are some of the agro sectors that have shown marked growth over the years. linkages between members of the food supply chains and prevailing policies and business environments to take advantage of the global market.

PROCESSED FOOD SEGMENT

The processing level of the agro industry may be at the primary, secondary or tertiary stage. In the case of hides and skins, India exports largely semi-processed items whereas in coffee/tea, the exports are mostly in secondary stage by way of fully processed bulk shipments without branding/packing. Exports at the tertiary stage mean branding and packaging the product that are ready for use by the consumer.

A few years ago, companies struggled to sell packaged foods. But now it is much easier to break into the Indian market because of a younger population, higher incomes, new technologies and a growing middle class, estimated at 50 million households. An average Indian spends around 53 per cent of his/her income on food. The domestic market for processed foods is not only huge but is growing fast in tandem with the economy. It is estimated to be worth $90 billion. Processed Food Manufacturing companies are required to be persistent and must adapt products to the Indian cultural preferences.

Many big companies like ITC, HLL, Nestle entered the Indian market a long time ago and have made a deep penetration in the market. From these success stories we can learn some lessons in order to capture the higher end of the local market and get a fair share of the export market. The model is structured around the following:-

- Large scale investment and adoption of the latest technologies
- Intensive marketing efforts
- Perhaps, a foreign tie-up can be beneficial
- Brand name.

The levels of processing and manufacturing can be classified into three groups, namely manual, mechanical and chemical or a combination thereof. In choosing the process, the main considerations are the nature of the raw materials, technology of processing, and packing.

OTHER SEGMENTS

Dairy product is another area where there is enormous potential. No doubt the country has made tremendous strides in the last 20 years in production and processing of milk and milk products. But the fact remains that only 15 per cent of all the milk produced is processed. Today, a large number of people suffer from diabetic or cardiac ailments and availability of fat free milk, fat free curd and sugar free food is poor. A simple product like soya milk is not produced in adequate quantity.

Fish and shrimp have good export potential but there is an immense lack of cold storage and modern processing facilities. For instance fish production is around six million tonnes a year and the frozen storage capacity spread over 500 units is only one lakh tonnes.

Another area is herbal medicine. It is being increasingly realized the world over that herbal drugs do not have any side effects. India has a good number of tried and tested herbal products in use and what is required is rigorous quality control, proper packaging and a brand name.

The government and modern retailers are addressing these issues with new laws on packaging and labeling as well as greater investment in the supply chain.

THE PROGRESS AHEAD

With modernization, innovation and incorporation of latest trends and technology in the entire food chain as well as agro-production, the total production capacity of agro products in India and the world is likely to double by the next decade.

India is the second largest producer of food in the world. Whether it is canned food, processed food, food grains, dairy products, frozen food, fish, meat, poultry, the Indian agro industry has a huge potential, the significance and growth of which will never cease.

Sea fishing, aqua culture, milk and milk products, meat and poultry are some of the agro sectors that have shown marked growth over the years.

INDIAN AGRICULTURAL INDUSTRY

Size of the Industry	It contributes almost 18.5% to India's GDP, while it accounts for about 13% of the country's total exports.
Geographical distribution	All over the country
Output per annum	According to the analysts the turnover of the total food market is approximately Rs.250000 crores out of which value-added food products comprises of Rs.80000 crores.
Market Capitalization	During 1990-91, the contribution of agriculture to the country's GDP output was at 32%, which was decreased to 20% by 2005-06.

HISTORY

Indian 'agro' or 'agriculture' marks the beginning of 'civilized' or 'sedentary' society. Climate change and increase in population in the country during the Holocene Era (10,000 BC onwards) led to the evolution of agriculture. Domestication of plants and animals during the Bronze Age transformed the profession of the early Homo sapiens from hunting and gathering to selective hunting, herding and finally to settled agriculture. Then gradually agricultural practices enabled people to establish permanent settlements and expand urban based societies in India.

According to history of the Indian agriculture industry the water-management practices are known to have either been taken up by the state or by local village communities since the earliest times. Regional rulers & the local representatives of the state were generally obliged to allocate a certain percentage of the agricultural taxes on building and managing water-storage, water-harvesting and/or water-diverting structures which facilitated a second crop, and provided water for drinking and other purposes in the long dry season.

The British rule in India witnessed the destruction of century-old water management structures and a virtual wreckage of the knowledge systems and cultural traditions that had helped build and preserve these water-management techniques over the centuries in states such as Bihar, Bengal, Karnataka, Tamil Nadu and others. Indian had a stagnant performance of agriculture in during the colonial period which turned into a sustained growth since 1947, then with a stronger performance in India especially in terms of per-capita food production the Indian Agricultural Industry grew.

After Independence the early years witnessed accentuation (emphasis) on the development of infrastructure for scientific agriculture. The Important steps were like including the establishment of fertilizer, pesticide factories, construction of large multi-purpose irrigation-cum-power projects, organization of community development projects and national extension programmes and even most majorly starting the agricultural universities as well as new agricultural research institutions across the length and breadth of the nation.

GREEN REVOLUTION

In Green Revolution the task Policy makers and planners took care about the concerns of national food independence, security, and political stability realized that self-sufficiency in food production was an absolute prerequisite. This particular perception led to a program of agricultural improvement called the Intensive Agriculture District Programme (IADP) and then eventually to the Green Revolution. The National Bank for Agriculture and Rural Development (NABARD) was set up for more easy transaction of money for the farmers all around the country. All these steps taken up by the Indian Government led to the quantum jump in the productivity and production of crops.

WHITE AND YELLOW REVOLUTION

The Green revolution in India generated a mood of self-confidence in the agricultural capability, which led to the next phase which was the Technology Mission. Under White and Yellow Revolution approach, the focus was mainly on conservation, cultivation, consumption, and commerce. An end-to-end approach was introduced which involved attention from all links in the production-consumption chain, by which the progress was steady and sometimes striking as in the case of milk and egg production.

PRESENT TIMES

Today Indian Agriculture Industry continues to face internal and external challenges. While the reasons might be monsoon dependence, low level of input usage, fragmented land-holding, lack of technology application, antiquated agronomic practices and poor rural infrastructure that would deter a healthy growth. While even the subsidies and barriers have been distorting international agricultural trade, rendering agri-exports from developing nations such as India uncompetitive.

BRIEF INTRODUCTION

The Indian Agriculture Industry is at the verge of a revolution that will modernize the entire food chain in India and as the total food production in India is likely to double in the next ten years. Today there are excellent export prospects, competitive pricing of agricultural product standards that are compared internationally which has created trade opportunities in the agro industry. This has enabled Indian Agriculture Industry Portal to serve as a main means by which every exporter and importer of India and abroad, could fulfill their requirements and avail the benefits of agro related buy-sell trade leads and other business opportunities.

This Indian Agricultural Industry revolution brings along with it the opportunities of profitable investment. The Indian government initiative has taken up to improve the Agricultural Standards by formulating the policies like EXIM policy, price policy, seed policy and this has paved the way for the profits to the Agricultural Industry in India. The statistics convey that the Indian Agricultural Industry today is the world's second largest producer of food. From canned, dairy, processed, frozen food to fisheries, meat, poultry, food grains, alcoholic beverages & soft drinks, the Indian agro industry has dainty areas to choose for business.

MARKET CAPITALIZATION

According to the analysts the turnover of the total food market is approximately Rs.250000 crores out of which value-added food products comprises of Rs.80000 crores. The Indian Government has also approved proposals for foreign collaborations, joint ventures, and industrial Licenses and 100% export oriented units predicting an investment of Rs.19100 crores out of which foreign investment is over Rs. 9100 crores.

Indian Agricultural industry also assumes significance owing to India's sizable agrarian economy, which contributes over 35% of GDP and employs around 65 % of the population. The consumer food segment goods have top priority both in terms of foreign investment and number of joint- ventures / foreign collaborations. Indian agricultural industry that have the capacity to lure foreigners with promising benefits are the deep sea fishing, aqua culture, milk and milk products, meat and poultry segments.

TOTAL CONTRIBUTION TO THE ECONOMY/ SALES

During 1990-91, the contribution of agriculture to the country's GDP output was at 32%, which was decreased to 20% by 2005-06. While the industry and services sectors contributed to 80% of the GDP which would increase further, with all of India's big corporate houses focusing on infrastructure and IT segments. The GDP contribution for sector depends on how much investments it can attract from the people, and the overdependence on industrial sectors would mean that India is slowly shifting from an agrarian based economy to that dependant on imports and exports.

The agricultural sector in India is the largest sector in the country's economy. It contributes almost 18.5% to India's GDP, while it accounts for about 13% of the country's total exports. For the past couple of years (2006-07 and 2007-08), the agriculture and allied sector saw a growth rate of more than 4%, comparing to the growth rate of 2.5% during the 10th Five Year Plan. Due to the huge market potential, Indian Agricultural Industry has become the home to a number of top agricultural companies.

TOP LEADING COMPANIES

- Monsanto
- Rallis

- H. J. Heinz
- Advanta India Ltd.
- Phalada Agro Research Foundation Ltd.
- Poabs Organic Estates
- National Agro Industries
- DuPont India
- Rasi Seeds
- ABT Industries

EMPLOYMENT OPPORTUNITIES

Over the years many significant studies have been made in agricultural production towards ensuring food security. Indian Agricultural Universities take care of the Agricultural Research and Education by the way of institutions and Postgraduate colleges for agricultural education and veterinary science. Agriculture has been regarded as the mainstay of our economy. It provides food to millions and raw material for our industries and as India also has enormous potential as an exporter of agricultural commodities. Indian variety of climatic types and natural conditions offer a conducive environment for the cultivation of a large variety of agricultural produce.

Work in agricultural sectors ranges from the academic pursuits of research and teaching to inspection, supervisory work and to other commercial activities of establishing farms, plantations, orchards and carrying on exports of produce from them. Basic training in agriculture is offered for Bachelor's degree course of three to four years duration at all the agricultural universities in the country. For graduate level courses candidates can take admission directly at any state agricultural university. The eligibility criterion for an individual would be 12th pass in science / agriculture. Many universities fix a minimum aggregate of 50% marks to be obtained in the qualifying examination.

Latest developments

- Establishing an enabling environment for agribusiness development - To work on this potential, the Indian government has taken several initiatives to encourage private investment in this area like deregulation of the sector from licenses, import duties and relaxing FDI norms, changing policy regulations for private sector activity in market. Initiatives to enhance the agricultural infrastructure by investing in Food Parks, Agri Export Zones, and terminal markets. All these initiatives are attached on a draft paper on creating an enabling environment for agribusiness development in India.
- As a consequence to the development initiatives taken up by the Indian government, many large Indian conglomerates are investing in agribusiness through by the way of agro-exports, processing, organized retailing and marketing. In recent years agricultural sector

has evidenced frantic activity because of huge investments by companies such as ITC, Bharati's Field Fresh, Reliance Industries Limited, PepsiCo, M&M's Mahindra Shubhlabh, and Aditya Birla Group among others.

- For Ensuring product quality and safety -A big challenge is faced by these Indian agro-industries is to ensure the availability of quantity and satisfactory quality of raw materials in order to meet the demands of processing, export and retail markets. Modern value chains are developed by these companies that are emerging alongside the traditional value chains to meet their specific quantity and quality requirements. Many research organizations have documented case studies of successful examples of modern supply chains where smallholder farmers in India are involved in meeting international quality standards. These draft case studies compare these farmers involved in modern value chains with their counterparts in the traditional markets from the perspective of farmer prices, risk management, access to inputs and production technology.

2

Paper

Paper is a thin material produced by pressing together moist fibres of cellulose pulp derived from wood, rags or grasses, and drying them into flexible sheets. It is a versatile material with many uses, including writing, printing, packaging, cleaning, and a number of industrial and construction processes.

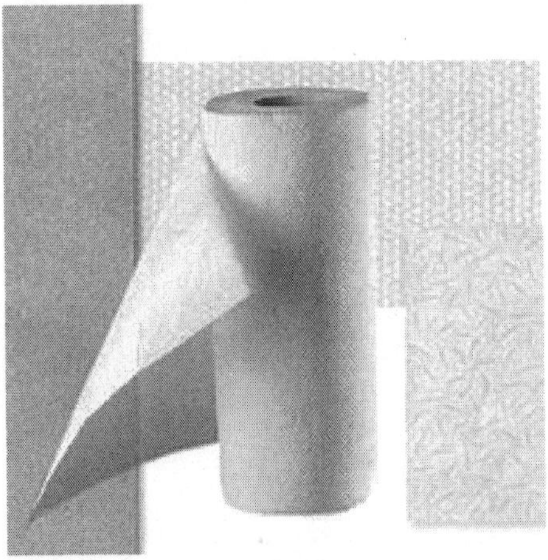

Fig. Different types of paper: carton,tissue paper

It and the pulp papermaking process is said to have been developed in China during the early 2nd century AD, possibly as early as the year 105 A.D., by the Han court eunuch Cai Lun, although the earliest archaeological fragments of paper derive from the 2nd century BC in China. The modern pulp and paper industry is global, with China leading its production and the United States right behind it.

HISTORY

The oldest known archaeological fragments of the immediate precursor to modern paper, date to the 2nd century BC in China. The pulp papermaking process is ascribed to Cai Lun, a 2nd-century AD Han court eunuch. With paper as an effective substitute for silk in many applications, China could export silk in greater quantity, contributing to a Golden Age.

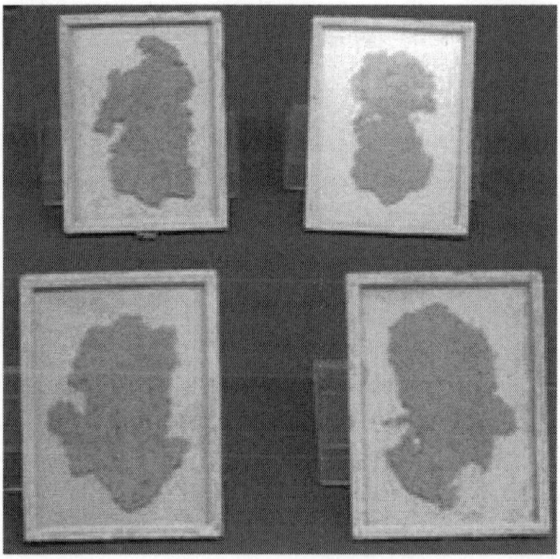

Fig. Hemp wrapping paper, China, circa 100 BC.

Its knowledge and uses spread from China through the Middle East to medieval Europe in the 13th century, where the first water powered paper mills were built. Because of paper's introduction to the West through the city of Baghdad, it was first called*bagdatikos*. In the 19th century, industrial manufacture greatly lowered its cost, enabling mass exchange of information and contributing to significant cultural shifts. In 1844, the Canadian inventor Charles Fenerty and the German F.G. Keller independently developed processes for pulping wood fibres.

ETYMOLOGY

The word "paper" is etymologically derived from Latin *papyrus*, which comes from the Greek ðÜðõñïò (*papuros*), the word for the*Cyperus papyrus* plant. Papyrus is a thick, paper-like material produced from the pith of the *Cyperus papyrus* plant which was used inancient Egypt and other Mediterranean cultures for writing before the introduction of paper into the Middle East and Europe. Although the word paper is etymologically derived from papyrus, the two are produced very differently and the development of the first is distinct from the development of the second. Papyrus is a lamination of natural plant fibres, while paper is manufactured from fibres whose properties have been changed by maceration.

PAPERMAKING

Chemical pulping

To make pulp from wood, a chemical pulping process separates lignin from cellulose fibres. This is accomplished by dissolving lignin in a cooking liquor, so that it may be washed from the cellulose; this preserves the length of the cellulose fibres. Paper made from chemical pulps are also known as wood-free papers–not to be confused with tree-free paper; this is because they do not contain lignin, which deteriorates over time. The pulp can also be bleached to produce white paper, but this consumes 5% of the fibres; chemical pulping processes are not used to make paper made from cotton, which is already 90% cellulose.

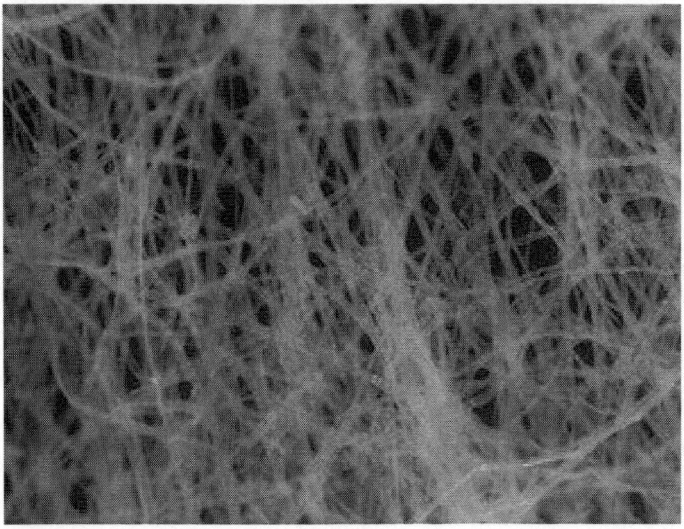

Fig. The microscopic structure of paper: Micrograph of paper autofluorescing under ultraviolet illumination. The individual fibres in this sample are around 10 μmin diameter.

There are three main chemical pulping processes: the sulfite process dates back to the 1840s and it was the dominant method extent before the second world war. The kraft process, invented in the 1870s and first used in the 1890s, is now the most commonly practiced strategy, one of its advantages is the chemical reaction with lignin, that produces heat which can be used to run a generator. Most pulping operations using the kraft process are net contributors to the electricity grid or use the electricity to run an adjacent paper mill. Another advantage is that this process recovers and reuses all inorganic chemical reagents. Soda pulping is another specialty process used to pulp straws, bagasse and hardwoods with high silicatecontent.

Mechanical pulping

There are two major mechanical pulps, the thermomechanical one (TMP) and groundwood pulp (GW). In the TMP process, wood is chipped and then fed into large steam heated refiners, where the chips are squeezed and converted to fibres between two steel discs. In the groundwood process, debarked logs are fed into grinders where they are pressed against rotating stones to be made into fibres. Mechanical pulping does not remove the lignin, so the yield is very high, >95%, however it causes the paper thus produced to turn yellow and become brittle over time. Mechanical pulps have rather short fibres, thus producing weak paper. Although large amounts of electrical energy are required to produce mechanical pulp, it costs less than the chemical kind.

De-inked pulp

Paper recycling processes can use either chemically or mechanically produced pulp; by mixing it with water and applying mechanical action the hydrogen bonds in the paper can be broken and fibres separated again. Most recycled paper contains a proportion of virgin fibre for the sake of quality; generally speaking, de-inked pulp is of the same quality or lower than the collected paper it was made from.

There are three main classifications of recycled fibre:.

- Mill broke or internal mill waste & nbsp – This incorporates any substandard or grade-change paper made within the paper mill itself, which then goes back into the manufacturing system to be re-pulped back into paper. Such out-of-specification paper is not sold and is therefore often not classified as genuine reclaimed recycled fibre, however most paper mills have been reusing their own waste fibre for many years, long before recycling become popular.
- Preconsumer waste – This is offcut and processing waste, such as guillotine trims and envelope blank waste; it is generated outside the paper mill and could potentially go to landfill, and is a genuine recycled fibre source; it includes de-inked preconsumer (recycled material that has been printed but did not reach its intended end use, such as waste from printers and unsold publications).
- Postconsumer waste – This is fibre from paper which has been used for its intended end use and would include office waste, magazine papers and newsprint. As the vast majority of this material has been printed - either digitally or by more conventional means such as lithography or rotogravure - it will either be recycled as printed paper or go through a de-inking process first.

Recycled papers can be made from 100% recycled materials or blended with virgin pulp, although they are (generally) not as strong nor as bright as papers made from the latter.

Additives

Besides the fibres, pulps may contain fillers such as chalk or china clay, which improve its characteristics for printing or writing. Additives for sizing purposes may be mixed with it and/or applied to the paper web later in the manufacturing process; the purpose of such sizing is to establish the correct level of surface absorbency to suit ink or paint.

Producing paper

The pulp is fed to a paper machine where it is formed as a paper web and the water is removed from it by pressing and drying.

Pressing the sheet removes the water by force; once the water is forced from the sheet, a special kind of felt, which is not to be confused with the traditional one, is used to collect the water; whereas when making paper by hand, a blotter sheet is used instead. Drying involves using air and/or heat to remove water from the paper sheets; in the earliest days of paper making this was done by hanging the sheets like laundry; in more modern times various forms of heated drying mechanisms are used. On the paper machine the most common is the steam heated can dryer. These can reach temperatures above 200 °F (93 °C) and are used in long sequences of more than 40 cans; where the heat produced by these can easily dry the paper to less than 6% moisture.

Finishing

The paper may then undergo sizing to alter its physical properties for use in various applications.

Paper at this point is *uncoated*. Coated paper has a thin layer of material such as calcium carbonate or china clay applied to one or both sides in order to create a surface more suitable for high-resolution halftone screens. (Uncoated papers are rarely suitable for screens above 150 lpi.) Coated or uncoated papers may have their surfaces polished bycalendering. Coated papers are divided into matte, semi-matte or silk, and gloss. Gloss papers give the highest optical density in the printed image.

The paper is then fed onto reels if it is to be used on web printing presses, or cut into sheets for other printing processes or other purposes. The fibres in the paper basically run in the machine direction. Sheets are usually cut "long-grain", i.e. with the grain parallel to the longer dimension of the sheet.

All paper produced by paper machines as the Fourdrinier Machine are wove paper, i.e. the wire mesh that transports the web leaves a pattern that has the same density along the paper grain and across the grain. Textured finishes, watermarks and wire patterns imitating hand-made *laid* paper can be created by the use of appropriate rollers in the later stages of the machine.

Wove paper does not exhibit "laidlines", which are small regular lines left behind on paper when it was handmade in a mould made from rows of metal wires or bamboo. Laidlines are very close together. They run perpendicular to the "chainlines", which are further apart. Handmade paper similarly exhibits "deckle edges", or rough and feathery borders.

APPLICATIONS

Paper can be produced with a wide variety of properties, depending on its intended use.

- For representing value: paper money, bank note, cheque, security, voucher and ticket
- For storing information: book, notebook, magazine, newspaper, art, zine, letter
- For personal use: diary, note to remind oneself, etc.; for temporary personal use: scratch paper
- For communication: between individuals and/or groups of people.
- For packaging: corrugated box, paper bag, envelope, Packing & Wrapping Paper, Paper string, Charta emporetica and wallpaper
- For cleaning: toilet paper, handkerchiefs, paper towels, facial tissue and cat litter
- For construction: papier-mâché, origami, paper planes, quilling, paper honeycomb, used as a core material in composite materials, paper engineering, construction paperand paper clothing
- For other uses: emery paper, sandpaper, blotting paper, litmus paper, universal indicator paper, paper chromatography, electrical insulation paper and filter paper

TYPES, THICKNESS AND WEIGHT

Fig. Card and paper stock for crafts use comes in a wide variety of textures and colors.

The thickness of paper is often measured by caliper, which is typically given in thousandths of an inch in the United States and in thousandths of a mm in the rest of the world. Paper may be between 0.07 and 0.18 millimetres (0.0028 and 0.0071 in) thick.

Paper is often characterized by weight. In the United States, the weight assigned to a paper is the weight of a ream, 500 sheets, of varying "basic sizes", before the paper is cut into the size it is sold to end customers. For example, a ream of 20 lb, 8.5 in × 11 in (216 mm × 279 mm) paper weighs 5 pounds, because it has been cut from a larger sheet into four pieces. In the United States, printing paper is generally 20 lb, 24 lb, or 32 lb at most. Cover stock is generally 68 lb, and 110 lb or more is considered card stock.

In Europe, and other regions using the ISO 216 paper sizing system, the weight is expressed in grammes per square metre (g/m^2 or usually just g) of the paper. Printing paper is generally between 60 g and 120 g. Anything heavier than 160 g is considered card. The weight of a ream therefore depends on the dimensions of the paper and its thickness.

Most commercial paper sold in North America is cut to standard paper sizes based on customary units and is defined by the length and width of a sheet of paper.

The ISO 216 system used in most other countries is based on the surface area of a sheet of paper, not on a sheet's width and length. It was first adopted in Germany in 1922 and generally spread as nations adopted the metric system. The largest standard size paper is A0 (A zero), measuring one square meter (approx. 1189 × 841 mm). Two sheets of A1, placed upright side by side fit exactly into one sheet of A0 laid on its side. Similarly, two sheets of A2 fit into one sheet of A1 and so forth. Common sizes used in the office and the home are A4 and A3 (A3 is the size of two A4 sheets).

The density of paper ranges from 250 kg/m^3 (16 lb/cu ft) for tissue paper to 1,500 kg/m^3 (94 lb/cu ft) for some speciality paper. Printing paper is about 800 kg/m^3 (50 lb/cu ft).

Paper may be classified into seven categories:

- *Printing papers* of wide variety.
- *Wrapping papers* for the protection of goods and merchandise. This includes wax and kraft papers.
- *Writing paper* suitable for stationery requirements. This includes ledger, bank, and bond paper.
- *Blotting papers* containing little or no size.
- *Drawing papers* usually with rough surfaces used by artists and designers, including cartridge paper.
- *Handmade papers* including most decorative papers, Ingres papers, Japanese paper and tissues, all characterized by lack of grain direction.
- *Specialty papers* including cigarette paper, toilet tissue, and other industrial papers.

Some paper types include:

- Bank paper
- Banana paper
- Bond paper
- Book paper
- Coated paper: glossy and matte surface

- Construction paper/sugar paper
- Cotton paper
- Fish paper (vulcanized fibres for electrical insulation)
- Inkjet paper
- Kraft paper
- Laid paper
- Leather paper
- Mummy paper
- Oak Tag Paper
- Sandpaper
- Tyvek paper
- Wallpaper
- Washi
- Waterproof paper
- Wax paper
- Wove paper
- Xuan paper

PAPER STABILITY

Much of the early paper made from wood pulp contained significant amounts of alum, a variety of aluminium sulfate salts that is significantly acidic. Alum was added to paper to assist in sizing, making it somewhat water resistant so that inks did not "run" or spread uncontrollably. Early papermakers did not realize that the alum they added liberally to cure almost every problem encountered in making their product would eventually be detrimental. The cellulose fibres that make up paper are hydrolyzed by acid, and the presence of alum would eventually degrade the fibres until the paper disintegrated in a process that has come to be known as "slow fire". Documents written on rag paper were significantly more stable. The use of non-acidic additives to make paper is becoming more prevalent, and the stability of these papers is less of an issue.

Paper made from mechanical pulp contains significant amounts of lignin, a major component in wood. In the presence of light and oxygen, lignin reacts to give yellow materials, which is why newsprint and other mechanical paper yellows with age. Paper made from bleached kraft or sulfite pulps does not contain significant amounts of lignin and is therefore better suited for books, documents and other applications where whiteness of the paper is essential.

Paper made from wood pulp is not necessarily less durable than a rag paper. The ageing behavior of a paper is determined by its manufacture, not the original source of the fibres. Furthermore, tests sponsored by the Library of Congress prove that all paper is at risk of acid decay, because cellulose itself produces formic, acetic, lactic and oxalic acids.

Mechanical pulping yields almost a tonne of pulp per tonne of dry wood used, which is why mechanical pulps are sometimes referred to as "high yield" pulps. With almost twice the yield as chemical pulping, mechanical pulps is often cheaper. Mass-market paperback books and newspapers tend to use mechanical papers. Book publishers tend to useacid-free paper, made from fully bleached chemical pulps for hardback and trade paperback books.

ENVIRONMENTAL IMPACT OF PAPER

The production and use of paper has a number of adverse effects on the environment.

Worldwide consumption of paper has risen by 400% in the past 40 years leading to increase in deforestation, with 35% of harvested trees being used for paper manufacture. Most paper companies also plant trees to help regrow forests. Logging of old growth forests accounts for less than 10% of wood pulp, but is one of the most controversial issues.

Paper waste accounts for up to 40% of total waste produced in the United States each year, which adds up to 71.6 million tons of paper waste per year in the United States alone. The average office worker in the US prints 31 pages every day. Americans also use on the order of 16 billion paper cups per year.

Conventional bleaching of wood pulp using elemental chlorine produces and releases into the environment large amounts of chlorinated organic compounds, including chlorinated dioxins. Dioxins are recognized as a persistent environmental pollutant, regulated internationally by the Stockholm Convention on Persistent Organic Pollutants. Dioxins are highly toxic, and health effects on humans include reproductive, developmental, immune and hormonal problems.

They are known to be carcinogenic. Over 90% of human exposure is through food, primarily meat, dairy, fish and shellfish, as dioxins accumulate in the food chain in the fatty tissue of animals.

FUTURE OF PAPER

Some manufacturers have started using a new, significantly more environmentally friendly alternative to expanded plastic packaging. Made out of paper, and known commercially as paperfoam, the new packaging has very similar mechanical properties to some expanded plastic packaging, but is biodegradable and can also be recycled with ordinary paper.

With increasing environmental concerns about synthetic coatings (such as PFOA) and the higher prices of hydrocarbon based petrochemicals, there is a focus on zein (corn protein) as a coating for paper in high grease applications such as popcorn bags.

Also, synthetics such as Tyvek and Teslin have been introduced as printing media as a more durable material than paper.

HISTORY OF PAPER

Paper was invented in ancient China during the Han dynasty (206 BC – 220 AD) and spread slowly to the west via the Silk Road.Papermaking and manufacturing in Europe was started by Muslims living on the Iberian Peninsula, (today's Portugal and Spain) andSicily in the 10th century, and slowly spread to Italy and Southern France reaching Germany by 1400. Earlier, other paper-like materials were in use including papyrus, parchment, palm leaves and vellum, but all of these were derived from materials which were expensive or in limited supply, or required extensive hand-processing to produce a satisfactory finish. Paper, being made from wood or rags, could be produced anywhere, and once large scale production techniques had been developed it could be manufactured in almost any quantity at moderate cost.

In medieval Europe, the hitherto handcraft of papermaking was mechanized by the use of waterpower, the first water papermill in the Iberian Peninsula having been built in the Portuguese city of Leiria in 1411, and other processes. The rapid expansion of European paper production was truly enhanced by the invention of the printing press and the beginning of the Printing Revolution in the 15th century.

The word "paper" is etymologically derived from *papyros*, Ancient Greek for the *Cyperus papyrus* plant. Papyrus is a thick, paper-like material produced from the pith of the *Cyperus papyrus* plant which was used in ancient Egypt and other Mediterranean cultures forwriting long before the making of paper in China. Papyrus however are plants dried and woven, while paper is manufactured from fibers whose properties have been changed by maceration or disintegration.

EARLY PAPERMAKING IN CHINA

Papermaking has traditionally been traced to China when Cai Lun, an official attached to the Imperial court during the Han Dynasty (202 BC-AD 220), created a sheet of paper using mulberry and other bast fibres along with fishnets, old rags, and hemp waste, though the earliest piece of paper found, at Fangmatan in Gansu province inscribed with a map, dates from 179-41 BC.

Techniques

During the Shang (1600–1050 BC) and Zhou (1050-256 BC) dynasties of ancient China, documents were ordinarily written on bone orbamboo (on tablets or on bamboo strips sewn and rolled together into scrolls), making them very heavy, awkward, and hard to transport. The light material of silk was sometimes used as a recording medium, but was normally too expensive to consider. The Han dynastyChinese court official Cai Lun (ca. 50–121) is widely regarded as the inventor of the modern method of papermaking (inspired by wasps and bees) from rags and other plant fibers) in 105. However, the discovery of specimens bearing written Chinese characters in 2006 atFangmatan in north-east China's Gansu Province suggest that paper was in use by the ancient Chinese military more than 100 years before Cai, in 8 BC, and possibly much earlier as the map fragment found at the Fangmatan tomb site dates from the early 2nd century BC. It therefore would appear that "Cai Lun's contribution was to improve this skill systematically and scientifically, fix a recipe for papermaking".

The record in the standard history says

In ancient times writings and inscriptions were generally made on tablets of bamboo or on pieces of silk called chih. But silk being costly and bamboos heavy they were not convenient to use. Tshai Lun then initiated the idea of making paper from the bark of trees, remnants of hemp, rags of cloth and fishing nets. He submitted the process to the emperor in the first year of Yuan-Hsing (105 AD) and received praise for his ability. From this time, paper has been in use everywhere and is universally called the paper of Marquis Tshai.

The manufacture may have originated from the practice of pounding and stirring rags in water, after which the matted fibres were collected on a mat. The bark of Paper Mulberry was particularly valued and high quality paper was developed in the late Han period, which used the bark of *tan* (j; sandalwood). In the Eastern Jin period paper began to be made on a fine bamboo screen-mould, treated with insecticidal dye for permanence. After printing became popular in the Song dynasty the demand grew more. Paper was often used as a levy, with one prefecture sending some 1.5 million sheets of paper to the capital as tribute up to the year 1101.

Uses

The first use of paper has been excavated in China dating to the reign of Emperor Wu of Han from the 2nd century BC, used for purposes of wrapping or padding protection for delicate bronze mirrors. It was also used for safety, such as the padding of poisonous 'medicine' as mentioned in the official history of the period. Although paper used for writing became widespread by the 3rd century AD, paper continued to be used for wrapping (and other) purposes. Toilet paper was used in China from around 875 C.E. In 589, the Chinese scholar-official Yan Zhitui (531-591) wrote: "Paper on which there are quotations or commentaries from Five Classics or the names of sages, I dare not use for toilet purposes". An Arab traveler to China once wrote of the curious Chinese tradition of toilet paper in 851, writing: "... [the Chinese] do not wash themselves with water when they have done their necessities; but they only wipe themselves with paper".

During the Tang dynasty (618–907) paper was folded and sewn into square bags to preserve the flavor of tea. During the same period, it was written that tea was served from baskets with multi-colored paper cups and paper napkins of different size and shape.During the Chinese Song dynasty (960–1279) not only did the government produce the world's first known paper-printed money, orbanknote, but paper money bestowed as gifts to deserving government officials were wrapped in special paperenvelopes.

DIFFUSION OF PAPER

After its origin in central China, the production and use of paper spread steadily. It is clear that paper was used at Dunhuang by AD 150, in Loulan in the modern-day province ofXinjiang by 200, and in Turpan by 399. Paper was concurrently introduced in Japan sometime between the years 280 and 610.

Islamic world

After the defeat of the Chinese in the Battle of Talas in 751 (present day Kyrgyzstan), the invention spread to the Middle East.

The legend goes, the secret of papermaking was obtained from two Chinese prisoners from the Battle of Talas, which led to the firstpaper mill in the Islamic world being founded in Samarkand. There are records of paper being made at Gilgit in Pakistan by the sixth century, in Samarkand in modern day Uzbekistan by 751, in Baghdad by 793, in Egypt by 900, and in Fes, Morocco around 1100

The laborious process of paper making was refined and machinery was designed for bulk manufacturing of paper. Production began inBaghdad, where a method was invented to make a thicker sheet of paper, which helped transform papermaking from an art into a major industry. The use of water-powered pulp mills for preparing the pulp material used in papermaking, dates back to Samarkand in the 8th century, though this should not be confused with paper mills. The Muslims also introduced the use of trip hammers (human- or animal-powered) in the production of paper, replacing the traditional Chinese mortar and pestle method. In turn, the trip hammer method was later employed by the Chinese.

By the 9th century, Muslims were using paper regularly, although for important works like copies of the revered Qur'an, vellum was still preferred. Advances in book production and bookbinding were introduced. In Muslim countries they made books lighter—sewn with silk and bound with leather-covered paste boards; they had a flap that wrapped the book up when not in use. As paper was less reactive to humidity, the heavy boards were not needed. By the 12th century in Marrakech in Morocco a street was named "Kutubiyyin" or book sellers which contained more than 100 bookshops.

The earliest recorded use of paper for packaging dates back to 1035, when a Persian traveler visiting markets in Cairo noted that vegetables, spices and hardware were wrapped in paper for the customers after they were sold.

Since the First Crusade in 1096, paper manufacturing in Damascus had been interrupted by wars, but its production continued in two other centres. Egypt continued with the thicker paper, while Iran became the center of the thinner papers. Papermaking was diffused across the Islamic world, from where it was diffused further west into Europe.Paper manufacture was introduced to India in the 13th century by Muslim merchants, where it almost wholly replaced traditional writing materials.

America

In America, archaeological evidence indicates that a similar bark-paper writing material was used by the Mayans no later than the 5th century AD. Called *amatl*, it was in widespread use among Mesoamerican cultures until the Spanish conquest. The paper is created by boiling and pounding the inner bark of trees, until the material becomes suitable for art and writing.

These materials made from pounded reeds and bark are technically not true paper, which is made from pulp, rags, and fibers of plants and cellulose.

Paper making as is more common to the European practice spread to the American continent first in Mexico by 1575 and then in Philadelphia by 1690.

Europe

The oldest known paper document in the West is the Mozarab Missal of Silos from the 11th century, probably using paper made in theIslamic part of the Iberian Peninsula. They used hemp and linen rags as a source of fiber. The first recorded paper mill in the Iberian Peninsula was in Xàtiva in 1151.

Papermaking reached Europe as early as 1085 in Toledo and was firmly established in Xàtiva, Spain by 1150. It is clear that France had a paper mill by 1190, and by 1276 mills were established in Fabriano, Italy and in Treviso and other northern Italian towns by 1340. Papermaking then spread further northwards, with evidence of paper being made in Troyes, France by 1348, in Holland sometime around 1340–1350, in Mainz, Germany in 1320, and in Nuremberg by 1390 in a mill set up by Ulman Stromer. This was just about the time when the woodcut printmaking technique was transferred from fabric to paper in the old master print and popular prints. There was a paper mill in Switzerland by 1432 and the first mill in England was set up by John Tate in 1490 near Stevenage in Hertfordshire, but the first commercially successful paper mill in Britain did not occur before 1588 when John Spilman set up a mill near Dartford in Kent.During this time, paper making spread to Poland by 1491, to Austria by 1498, to Russia by 1576, to the Netherlands by 1586, to Denmark by 1596, and to Sweden by 1612.

Arab prisoners who settled in a town called Borgo_Saraceno introduced Fabriano artisans to the technique of making paper by hand. At the time they were renowned for their wool-weaving and manufacture of cloth. Fabriano papermakers considered the process of making paper by hand an art form and were able to refine the process to successfully compete with parchment which was the primary medium for writing at the time. They developed the application of stamping hammers to reduce rags to pulp for making paper, sizing paper by means of animal glue, and creating watermarks in the paper during its forming process.

The Fabriano used glue, obtained by boiling scrolls or scraps of animal skin to size the paper; it is suggested that this technique was recommended by the local tanneries.

The introduction of the first European watermarks in Fabriano was linked to applying metal wires on a cover laid against the mould which was used for forming the paper.

They adapted the water wheels from the fuller's mills to drive a series of 3 wooden hammers per trough. The hammers were raised by their heads by cams fixed to a waterwheel's axle made from a large tree trunk.

Paper mills

A paper mill is a water-powered mill that pounds the pulp by the use of trip-hammers. The mechanization of the pounding process was an important improvement in paper manufacture over the manual pounding with hand pestles.

While the use of human and animal powered mills were known to Chinese and Muslim papermakers, evidence for water-powered paper mills is elusive in both of them. The general absence of the use of water-power in Muslim papermaking is suggested by the habit of Muslim authors to call a production center not a "mill", but a "paper manufactory".

Donald Hill has identified a possible reference to a water-powered paper mill in Samarkand, in the 11th-century work of the Persian scholar Abu Rayhan Biruni, but concludes that the passage is "too brief to enable us to say with certainty" that it refers to a water-powered paper mill. While this is seen by Halevi nonetheless as evidence of Samarkand first harnessing waterpower in the production of paper, he concedes that it is not known if waterpower was applied to papermaking elsewhere across the Islamic world at the time;Burns remains altogether sceptical given the isolated occurrence of the reference and the prevalence of manual labour in Islamic papermaking elsewhere.

The earliest certain evidence to a water-powered paper mill dates to 1282 in the Spanish Kingdom of Aragon. A decree by the Christian king Peter III addresses the establishment of a royal "molendinum", a proper hydraulic mill, in the paper manufacturing centre of Xàtiva.The crown innovation appears to be resented by the local Muslim paper makering community; the document guarantees the Muslim subjects the right to continue their way of traditional paper making by beating the pulp manually and grants them the right to be exempted from work in the new mill. Paper making centers began to multiply in the late 13th century in Italy, reducing the price of paper to one sixth of parchment and then falling further; paper making centers reached Germany a century later.

The first paper mill north of the Alps was established in Nuremberg by Ulman Stromer in 1390; it is later depicted in the lavishly illustrated *Nuremberg Chronicle*. From the mid-14th century onwards, European paper milling underwent a rapid improvement of many work processes.

Fiber sources

Before the industrialisation of the paper production the most common fibre source was recycled fibres from used textiles, called rags. The rags were from hemp, linen andcotton. A process for removing printing inks from recycled paper was invented by German jurist Justus Claproth in 1774. Today this method is called deinking. It was not until the introduction of wood pulp in 1843 that paper production was not dependent on recycled materials from ragpickers.

19TH CENTURY ADVANCES IN PAPERMAKING

Although cheaper than vellum, paper remained expensive, at least in book-sized quantities, through the centuries, until the advent of steam-driven paper making machines in the 19th century, which could make paper with fibres from wood pulp. Although older machines predated it, the Fourdrinier papermaking machine became the basis for most modern papermaking. Nicholas Louis Robert of Essonnes, France, was granted a patent for a continuous paper making machine in 1799.

At the time he was working for Leger Didot with whom he quarrelled over the ownership of the invention. Didot sent his brother-in-law, John Gamble, to meet Sealy and Henry Fourdrinier, stationers of London, who agreed to finance the project. Gamble was granted British patent 2487 on 20 October 1801. With the help particularly of Bryan Donkin, a skilled and ingenious mechanic, an improved version of the Robert original was installed at Frogmore, Hertfordshire, in 1803, followed by another in 1804. A third machine was installed at the Fourdriniers' own mill at Two Waters. The Fourdriniers also bought a mill at St Neots intending to install two machines there and the process and machines continued to develop.

However, experiments with wood showed no real results in the late 18th century and at the start of the 19th century. By 1800, Matthias Koops (in London, England) further investigated the idea of using wood to make paper, and in 1801 he wrote and published a book titled *Historical account of the substances which have been used to describe events, and to convey ideas, from the earliest date, to the invention of paper.* His book was printed on paper made from wood shavings (and adhered together). No pages were fabricated using the pulping method (from either rags or wood). He received financial support from the royal family to make his printing machines and acquire the materials and infrastructure need to start his printing business. But his enterprise was short lived. Only a few years following his first and only printed book (the one he wrote and printed), he went bankrupt. The book was very well done (strong and had a fine appearance), but it was very costly.

Then in the 1830s and 1840s, two men on two different continents took up the challenge, but from a totally new perspective. Both Friedrich Gottlob Keller and Charles Fenertybegan experiments with wood but using the same technique used in paper making; instead of pulping rags, they thought about pulping wood. And at about the same time, by mid-1844, they announced their findings. They invented a machine which extracted the fibres from wood (exactly as with rags) and made paper from it. Charles Fenerty also bleached the pulp so that the paper was white. This started a new era for paper making. By the end of the 19th-century almost all printers in the western world were using wood in lieu of rags to make paper.

Together with the invention of the practical fountain pen and the mass-produced pencil of the same period, and in conjunction with the advent of the steam driven rotary printing press, wood based paper caused a major transformation of the 19th century economy and society in industrialized countries. With the introduction of cheaper paper, schoolbooks, fiction, non-fiction, and newspapers became gradually available by 1900. Cheap wood based paper also meant that keeping personal diaries or writing letters became possible and so, by 1850, the clerk, or writer, ceased to be a high-status job.

The original wood-based paper was acidic due to the use of alum and more prone to disintegrate over time, through processes known as slow fires. Documents written on more expensive rag paper were more stable. Mass-market paperback books still use these cheaper mechanical papers, but book publishers can now use acid-free paperfor hardback and trade paperback books.

DETERMINING PROVENANCE

Determining the provenance of paper is a complex process that can be done in a variety of ways. The easiest way is using a known sheet of paper as an exemplar. Using known sheets can produce an exact identification. Next, comparing watermarks with those contained in catalogs or trade listings can yield useful results. Inspecting the surface can also determine age and location by looking for distinct marks from the production process. Chemical and fiber analysis can be used to establish date of creation and perhaps location.

PAPYRUS

The word papyrus /pY□paj□rY□s/ refers to a thick paper-like material made from the pith of the papyrus plant, *Cyperus papyrus*. *Papyrus*can also refer to a document written on sheets of papyrus joined together side by side and rolled up into a scroll, an early form of a book. The plural for such documents is papyri.

Papyrus is first known to have been used in ancient Egypt (at least as far back as the First Dynasty), as the *Cyperus papyrus* plant was a wetland sedge that was once abundant in the Sudd of Southern Sudan along with the Nile Delta of Egypt. Papyrus was also used throughout the Mediterranean region and in Kingdom of Kush. The Ancient Egyptians used papyrus as a writing material, as well as employing it commonly in the construction of other artifacts such as reed boats, mats, rope, sandals, and baskets.

History

Papyrus was first manufactured in Egypt and Southern Sudan as far back as the fourth millennium BCE. The earliest archaeological evidence of papyrus was excavated in 2012 and 2013 at Wadi al-Jarf, an ancient Egyptian harbor located on the Red Seacoast. These documents date from c. 2560–2550 BCE (end of the reign of Khufu). The papyrus rolls describe the last years of building the Great Pyramid of Giza. In the first centuries BCE and CE, papyrus scrolls gained a rival as a writing surface in the form ofparchment, which was prepared from animal skins. Sheets of parchment were folded to form quires from which book-form codiceswere fashioned. Early Christian writers soon adopted the codex form, and in the Græco-Roman world, it became common to cut sheets from papyrus rolls to form codices.

Codices were an improvement on the papyrus scroll, as the papyrus was not pliable enough to fold without cracking and a long roll, or scroll, was required to create large-volume texts. Papyrus had the advantage of being relatively cheap and easy to produce, but it was fragile and susceptible to both moisture and excessive dryness. Unless the papyrus was of perfect quality, the writing surface was irregular, and the range of media that could be used was also limited.

Papyrus was replaced in Europe by the cheaper, locally produced products parchment and vellum, of significantly higher durability in moist climates, though Henri Pirenne's connection of its disappearance with the Muslim conquest of Egypt is contested. Its last appearance in the Merovingian chancery is with a document of 692, though it was known in Gaul until the middle of the following century.

The latest certain dates for the use of papyrus are 1057 for a papal decree (typically conservative, all papal bulls were on papyrus until 1022), under Pope Victor II, and 1087 for an Arabic document. Its use in Egypt continued until it was replaced by more inexpensive paper introduced by Arabs. By the 12th century, parchment and paper were in use in the Byzantine Empire, but papyrus was still an option.

Papyrus was made in several qualities and prices; these are listed, with minor differences, both by Pliny and Isidore of Seville.

Until the middle of the 19th century, only some isolated documents written on papyrus were known. They did not contain literary works. The first discovery of papyri rolls in modern days was made at Herculaneum in 1752. Before that date, the only papyri known were a few surviving from medieval times.

Etymology

The English word "papyrus" derives, via Latin, from Greek ðÜðõñïò (*papuros*), a loanword of unknown (perhaps Pre-Greek) origin. Greek has a second word for it, âýâëïò (*bublos*, said to derive from the name of the Phoenician city of Byblos). The Greek writer Theophrastus, who flourished during the 4th century BC, uses *papuros* when referring to the plant used as a foodstuff and *bublos* for the same plant when used for nonfood products, such as cordage, basketry, or writing surfaces. The more specific term âßâëïò *biblos*, which finds its way into English in such words as 'bibliography', 'bibliophile', and 'bible', refers to the inner bark of the papyrus plant. *Papyrus* is also the etymon of 'paper', a similar substance.

Documents written on papyrus

The word for the material papyrus is also used to designate documents written on sheets of it, often rolled up into scrolls. The plural for such documents is papyri. Historical papyri are given identifying names—generally the name of the discoverer, first owner or institution where they are kept—and numbered, such as "Papyrus Harris I". Often an abbreviated form is used, such as "pHarris I". These documents provide important information on ancient writings; they give us the only extant copy of Menander, the Egyptian Book of the Dead, Egyptian treatises on medicine (the Ebers Papyrus) and on surgery (the Edwin Smith papyrus), Egyptian mathematical treatises (the Rhind papyrus), and Egyptian folk tales (theWestcar papyrus). When, in the 18th century, a library of ancient papyri was found in Herculaneum, ripples of expectation spread among the learned men of the time. However, since these papyri were badly charred, their unscrolling and deciphering is still going on today.

Manufacture and use

Papyrus is made from the stem of the papyrus plant, *Cyperus papyrus*. The outer rind is first removed, and the sticky fibrous inner pith is cut lengthwise into thin strips of about 40 cm (16 in) long. The strips are then placed side by side on a hard surface with their edges slightly overlapping, and then another layer of strips is laid on top at a right angle. The strips may have been soaked in water long enough fordecomposition to begin, perhaps increasing adhesion, but this is not certain. The two layers possibly were glued together. While still moist, the two layers are hammered together, mashing the layers into a single sheet. The sheet is then dried under pressure. After drying, the sheet is polished with some rounded object, possibly a stone or seashell or round hardwood.

Sheets could be cut to fit the obligatory size or glued together to create a longer roll. A wooden stick would be attached to the last sheet in a roll, making it easier to handle. To form the long strip scrolls required, a number of such sheets were united, placed so all the horizontal fibres parallel with the roll's length were on one side and all the vertical fibres on the other. Normally, texts were first written on the *recto*, the lines following the fibres, parallel to the long edges of the scroll. Secondarily, papyrus was often reused, writing across the fibres on the *verso*. Pliny the Elder describes the methods of preparing papyrus in his *Naturalis Historia*.

In a dry climate, like that of Egypt, papyrus is stable, formed as it is of highly rot-resistant cellulose; but storage in humid conditions can result inmolds attacking and destroying the material. Library papyrus rolls were stored in wooden boxes and chests made in the form of statues. Papyrus scrolls were organized according to subject or author, and identified with clay labels that specified their contents without having to unroll the scroll. In European conditions, papyrus seems to have lasted only a matter of decades; a 200-year-old papyrus was considered extraordinary.

Imported papyrus once commonplace in Greece and Italy has since deteriorated beyond repair, but papyrus is still being found in Egypt; extraordinary examples include the Elephantine papyri and the famous finds at Oxyrhynchus and Nag Hammadi. The Villa of the Papyri atHerculaneum, containing the library of Lucius Calpurnius Piso Caesoninus, Julius Caesar's father-in-law, was preserved by the eruption of Mount Vesuvius, but has only been partially excavated.

Sporadic attempts to revive the manufacture of papyrus have been made since the mid-18th century. Scottish explorer James Bruceexperimented in the late 18th century with papyrus plants from the Sudan, for papyrus had become extinct in Egypt. Also in the 18th century,Sicilian Saverio Landolina manufactured papyrus at Syracuse, where papyrus plants had continued to grow in the wild. During the 1920s, when Egyptologist Battiscombe Gunn lived in Maadi, outside Cairo, he experimented with the manufacture of papyrus, growing the plant in his garden.

He beat the sliced papyrus stalks between two layers of linen, and produced successful examples of papyrus, one of which was exhibited in theEgyptian Museum in Cairo.

The modern technique of papyrus production used in Egypt for the tourist trade was developed in 1962 by the Egyptian engineer Hassan Ragab using plants that had been reintroduced into Egypt in 1872 from France. Both Sicily and Egypt have centres of limited papyrus production.

Papyrus is still used by communities living in the vicinity of swamps, to the extent that rural householders derive up to 75% of their income from swamp goods. Particularly in East and Central Africa, people harvest papyrus, which is used to manufacture items that are sold or used locally. Examples include baskets, hats, fish traps, trays or winnowing mats and floor mats. Papyrus is also used to make roofs, ceilings, rope and fences. Although alternatives, such as eucalyptus, are increasingly available, papyrus is still used as fuel.

PAPERMAKING

Papermaking is the process of making paper, a material which is used universally today for writing and packaging.

In papermaking, a dilute suspension of fibres in water is drained through a screen, so that a mat of randomly interwoven fibres is laid down. Water is removed from this mat of fibres by pressing and drying to make paper. Since the invention of the Fourdrinier machine in the 19th century, most paper has been made from wood pulp because of cost. But other fibre sources such as cotton and textiles are used for high-quality papers. One common measure of a paper's quality is its non-wood-pulp content, e.g., 25% cotton, 50% rag, etc. Previously, paper was made up of rags and hemp as well as other materials.

HISTORY

Papermaking is known to have been traced back to China about 105 CE, when Cai Lun, an official attached to the Imperial court during the Han Dynasty (202 BCE-220 CE), created a sheet of paper using mulberry and other bast fibres along with fishnets, old rags, andhemp waste. However a recent archaeological discovery has been reported from Gansu province of paper with legible Chinese writings on it dating from 8 BCE, while paper had been used in China for wrapping and padding since the 2nd century BCE. Paper used as a writing medium became widespread by the 3rd century and, by the 6th century, toilet paper was starting to be used in China as well.During the Tang Dynasty (618-907 CE) paper was folded and sewn into square bags to preserve the flavour of tea, while the later Song Dynasty (960-1279 CE) was the first government to issue paper-printed money.

In the 8th century, paper spread to the Islamic world, where the rudimentary and laborious process of papermaking was refined and machinery was designed for bulk manufacturing of paper. Production began in Samarkand, Baghdad, Damascus, Cairo, Morocco thenMuslim Spain. In Baghdad, this was under the supervision of the Grand Vizier Ja'far ibn Yahya. In general, Muslims invented a method to make a thicker sheet of paper. This helped transform papermaking from an art into a major industry. The earliest use of water-powered mills in paper production, specifically the use of pulp mills for preparing the pulp for papermaking, dates back to Samarkand in the 8th century. The earliest references to paper mills also come from the medieval Islamic world, where they were first noted in the 9th century by Arabic geographers in Damascus. Papermaking was diffused across the Islamic world, from where it was diffused further west into Europe.

Modern papermaking began in the early 19th century in Europe with the development of the Fourdrinier machine, which produces a continuous roll of paper rather than individual sheets. These machines are considerably large, up to 150 meters in length, produce up to 10 meters wide sheet, and running around

100 km/h. In 1844, Canadianinventor Charles Fenerty and German inventor F.G. Keller had invented the machine and associated process to make use of wood pulp in papermaking. This would end the nearly 2,000-year use of pulped rags and start a new era for the production of newsprint and eventually almost all paper was made out of pulped wood.

MANUAL PAPERMAKING

Papermaking, regardless of the scale on which it is done, involves making a dilute suspension of fibres in water and allowing this suspension to drain through a screen, so that a mat of randomly interwoven fibres is laid down. Water is then removed from this mat of fibres using a press.

The method of manual papermaking changed very little over time, despite advances in technologies. The process of manufacturing handmade paper can be generalized into five steps:

1. Separating the useful fibre from the rest of raw materials. (e.g. cellulose from wood, cotton, etc.)
2. Beating down the fibre into pulp

Fig. An illustration from 105 AD depicting the papermaking process as designed by Cai Lun.

3. Adjusting the colour, mechanical, chemical, biological, and other properties of the paper by adding special chemical premixes

4. Screening the resulting solution
5. Pressing and drying to get the actual paper

Screening the fibre involves using a mesh made from non-corroding and inert material, such as aluminium, which is stretched in a wooden frame similar to that of a window.

The size of the paper is governed by the size of the frame. This tool is then completely submerged in the solution vertically and drawn out horizontally to ensure a uniform coating of the wire mesh. Excess water is then removed and the wet mat of fibre laid on top of a damp cloth. The process is repeated for the required number of sheets. This stack of wet mats is then pressed in a hydraulic press very gently to ensure the fibre does not squeeze out. The fairly damp fibre is then dried using a variety of methods, such as vacuum drying or simply air drying. Sometimes, the individual sheet is rolled to flatten, harden, and refine the surface. Finally, the paper is then cut to the desired shape or the standard shape (A4, letter, legal, etc.) and packed.

The wooden frame is called a "deckle". The deckle leaves the edges of the paper slightly irregular and wavy, called "deckle edges", one of the indications that the paper was made by hand. Deckle-edged paper is occasionally mechanically imitated today to create the impression of old-fashioned luxury. The impressions in paper caused by the wires in the screen that run sideways are called "laid lines" and the impressions made, usually from top to bottom, by the wires holding the sideways wires together are called "chain lines". Watermarks are created by weaving a design into the wires in the mould. This is essentially true of Oriental moulds made of other substances, such as bamboo. Handmade paper generally folds and tears more evenly along the laid lines.

Handmade paper is also prepared in laboratories to study papermaking and in paper mills to check the quality of the production process. The "handsheets" made according to TAPPI Standard T 205 are circular sheets 15.9 cm (6.25 in) in diameter and are tested for paper characteristics such as brightness, strength and degree of sizing.

INDUSTRIAL PAPERMAKING

A modern paper mill is divided into several sections, roughly corresponding to the processes involved in making handmade paper. Pulp is refined and mixed in water with other additives to make a pulp slurry. The head-box of the paper machine (Fourdrinier machine) distributes the slurry onto a moving continuous screen, water drains from the slurry (by gravity or under vacuum), the wet paper sheet goes through presses and dries, and finally rolls into large rolls. The outcome often weighs several tons.

Another type of paper machine makes use of a cylinder mould that rotates while partially immersed in a vat of dilute pulp. The pulp is picked up by the wire and covers the mould as it rises out of the vat. A couch roller is pressed against the mould to smooth out the pulp, and picks the wet sheet off the mould.

Key inventors include Henry Fourdrinier, Heinrich Voelter and Carl Daniel Ekman, among others.

HISTORY OF PAPER SHEET PRODUCTION

Folio

In the beginning of Western papermaking, paper size was fairly standard. A page of paper is referred to as a *leaf*. When a leaf was printed on without being folded, the size was referred to as *folio* (meaning leaf). It was roughly equal to the size of a small newspaper sheet.

Quarto

A Folio folded once produces two leaves (or four pages), and the size of these leaves was referred to as *quarto* (4to) (about 230 x 280 mm).

Octavo

If the original sheet was folded in half again, the result was eight pages, referred to as *octavo* (8vo), which is roughly the size of an average modern novel. An octavo folding produces four leaves; the first two and the second two will be joined at the top by the first fold. The top edge is usually *trimmed* to make it possible to look freely at each side of the leaf. Sometimes books are found that have not been trimmed on the top, and these pages are referred to as *unopened*.

An octavo book produces a printing puzzle. The paper was first printed before folding and thus pages 8 and 1 are printed right-side-up on the bottom of the sheet, and pages 4 and 5 are printed upside-down on the top of the same side of the paper. On the opposite side, pages 2 and 7 are printed right-side-up on the bottom of the sheet, and pages 6 and 3 are printed upside-down on the top of the sheet. When the paper is folded twice and the folds trimmed, the pages fall into proper order.

Sixteen-mo

Smaller books are produced by folding the leaves again to produce 16 pages, known as a *sixteen-mo* (16mo) (originally sextodecimo). Other folding arrangements produce yet smaller books such as the thirty-two-mo (32mo) (duo et tricensimo).

Octavo bookbinding

When a standard-sized octavo book is produced by twice folding a large leaf, two leaves joined at the top will be contained in the resulting fold (which ends up in the gulley between the pages). This group of eight numberable pages is called a *signature* or a *gathering*. Traditionally, printed signatures were stacked on top of each other in a *sewing frame* and each signature was sewn through the inner fold to the signature on top of it. The sewing ran around *leather bands* or *fabric tapes* along the backs of the signatures to stabilize the pile of signatures. The leather bands originally used in the West to stabilize the backs of sewn books appear as a number of *ridges* under the leather on the spine of leather books. The ends of the leather strips or fabric bands were sewn or glued onto the *cover boards* and reinforced the hinging of the book in its covers.

Standardisation ISO sizes

While opinions and speculation abound on exact reasons for standardized paper sizes, the most revealing feature of popular sizes (such as Letter and ISO 216 sizes) is that they conform not to some arbitrary device dimension, but that the length of the paper is chosen to be the width of the page times the square root of two (H"1.414). This feature allowed for a large page to be cut in half and the resulting two pages to have the same aspect ratio as the original piece (just with half the size). Repeated cuts can be made to reduce the entire sheet to one size of pages without wasted paper. This format was formalized by ISO 216 however such logic dictated efficient paper sizes long before the ISO was created. For example, traditional 8.5"x11" Letter paper is within a few millimetres of A4 paper (ISO 216) dimensions. While paper sizes "may" have been chosen based on the size of original frames, the frames themselves were chosen to make page reduction efficient without distorting the aspect ratio of the pages regardless of final size chosen. However, there are paper sizes that do not conform to this idea when specific applications are needed.

Vatmen paper

Vatmen Paper was a type of paper made in the Netherlands that was 17 inches (430 mm) wide and 44 inches (1,100 mm) long. 44 inches was (reputedly) chosen because that is how far the papermaker could stretch his arm. A single vatman can generally handle a mould and deckle which produce up to a 25 inches (640 mm) wide sheet.

Handmade paper

Kalpi is one of the important hubs of handmade paper industry in India. A layman can very well confuse the paper sheet as leather when it is dried in the sun.

PAPER MACHINE

Many modern papermaking machines are based on the principles of the Fourdrinier Machine, which uses a specially woven plastic fabric mesh conveyor belt (known as a *wire* as it was once woven from bronze) in the forming section, where a slurry of fibre (usually wood or other vegetable fibres) is drained to create a continuous paper web. After the forming section the wet web passes through a press section to squeeze out excess water, then the pressed web passes through a heated drying section.

The original Fourdrinier forming section used a horizontal drainage area, referred to as the *drainage table*.

Paper machines have four distinct operational sections:

- Forming section, commonly called the wet end, is where the slurry of fibres filters out fluid a continuous fabric loop to form a wet web of fibre.
- Press section where the wet fibre web passes between large rolls loaded under high pressure to squeeze out as much water as possible.
- Drying section, where the pressed sheet passes partly around, in a serpentine manner, a series of steam heated drying cylinders. Drying removes the water content down to a level of about 6%, where it will remain at typical indoor atmospheric conditions.
- Calender section where the dried paper is smoothened under high loading and pressure. Only one *nip* (where the sheet is pressed between two rolls) is necessary in order to hold the sheet, which shrinks through the drying section and is held in tension between the press section (or breaker stack if used) and the calender. Extra nips give more smoothing but at some expense to paper strength.

Paper machines are long-lived assets that usually remain in service for several decades. It is common to rebuild machines periodically to increase production and improve quality or to change the paper grade.

HISTORY OF PAPER MACHINES

long Before the invention of continuous paper making, paper was made in individual sheets by stirring a container of pulp slurry and either pouring it into a fabric sieve called a sheet mould or dipping and lifting the sheet mould from the vat. While still on the fabric in the sheet mould the wet paper is pressed to remove excess water and then the sheet was lifted off to be hung over a rope or wooden rod to air dry.

In 1799, Louis-Nicolas Robert of Essonnes, France, was granted a patent for a continuous paper making machine. At the time Robert was working for Saint-Léger Didot, with whom he quarrelled over the ownership of the invention. Didot thought that England was a better place to develop the machine. But during the troubled times of the French Revolution, he could not go there himself, so he sent his brother in law, John Gamble, an Englishman living in Paris. Through a chain of acquaintances, Gamble was introduced to the brothers Sealy and Henry Fourdrinier, stationers of London, who agreed to finance the project. Gamble was granted British patent 2487 on 20 October 1801. With the help particularly of Bryan Donkin, a skilled and ingenious mechanic, an improved version of the Robert original was installed at Frogmore Mill, Apsley, Hertfordshire, in 1803, followed by another in 1804. A third machine was installed at the Fourdriniers' own mill at Two Waters. The Fourdriniers also bought a mill at St Neots intending to install two machines there and the process and machines continued to develop.

Thomas Gilpin is most often credited for creating the first U.S cylinder type papermaking machine at Brandywine Creek, Delaware in 1817. This machine was also developed in England, but it was a cylinder mould machine. The Fourdrinier machine wasn't introduced into the USA until 1827.

However, records show Charles Kinsey of Patterson, NJ had already patented a continuous process papermaking machine in 1807. Kinsey's machine was built locally by Daniel Sawn and by 1809 the Kinsey machine was successfully making paper at the Essex Mill in Paterson. Financial stress and potential opportunities created by the Embargo of 1807eventually persuaded Kinsey and his backers to change the mill's focus from paper to cotton and Kinsey's early papermaking successes were soon overlooked and forgotten.

Gilpin's 1817 patent was similar to Kinsey's, as was the John Ames patent of 1822. The Ames patent was challenged by his competitors, asserting that Kinsey was the original inventor and Ames had been pilfering other peoples' ideas, their evidence being the employment of Daniel Sawn to work on his machine.

The method of continuous production demonstrated by the paper machine influenced the development of continuous rolling of iron and later steel and other continuous production processes.

PULP TYPES AND THEIR PREPARATIONS

The plant fibres used for pulp are composed mostly of cellulose and hemi-cellulose, which have a tendency to form molecular linkages between fibres in the presence of water. After the water evaporates the fibres remain bonded. It is not necessary to add additional binders for most paper grades, although both wet and dry strength additives may be added.

Rags of cotton and linen were the major source of pulp for paper before wood pulp. Today almost all pulp is of wood fibre. Cotton fibre is used in speciality grades, usually in printing paper for such things as resumes and currency.

Sources of rags often appear as waste from other manufacturing such as denim fragments or glove cuts. Fibres from clothing come from the cotton boll. The fibres can range from 3 to 7 cm in length as they exist in the cotton field. Bleach and other chemicals remove the colour from the fabric in a process of cooking, usually with steam. The cloth fragments mechanically abrade into fibres, and the fibres get shortened to a length appropriate for manufacturing paper with a cutting process. Rags and water dump into a trough forming a closed loop. A cylinder with cutting edges, or knives, and a knife bed is part of the loop. The spinning cylinder pushes the contents of the trough around repeatedly. As it lowers slowly over a period of hours, it breaks the rags up into fibres, and cuts the fibres to the desired length. The cutting process terminates when the mix has passed the cylinder enough times at the programmed final clearance of the knives and bed.

Another source of cotton fibre comes from the cotton ginning process. The seeds remain, surrounded by short fibres known as linters for their short length and resemblance to lint. Linters are too short for successful use in fabric. Linters removed from the cotton seeds are available as first and second cuts. The first cuts are longer.

The two major classifications of pulp are chemical and mechanical. Chemical pulps formerly used a sulphite process, but the kraft process is now predominant. Kraft pulp has superior strength to sulphite and mechanical pulps. Both chemical pulps and mechanical pulps may be bleached to a high brightness.

Chemical pulping dissolves the lignin that bonds fibres to one another, and binds the outer fibrils that compose individual fibres to the fibre core. Lignin, like most other substances that can separate fibres from one another, acts as a debonding agent, lowering strength. Strength also depends on maintaining long cellulose molecule chains.

The kraft process, due to the alkali and sulphur compounds used, tends to minimize attack on the cellulose and the non-crystalline hemicellulose, which promotes bonding, while dissolving the lignin. Acidic pulping processes shorten the cellulose chains.

Kraft pulp makes superior linerboard and excellent printing and writing papers.

Groundwood, the main ingredient used in newsprint and a principal component of magazine papers (coated publications), is literally ground wood produced by a grinder. Therefore it contains a lot of lignin, which lowers its strength. The grinding produces very short fibres that drain slowly.

Thermomechanical pulp (TMP) is a variation of groundwood where fibres are separated mechanically while at high enough temperatures to soften the lignin.

Between chemical and mechanical pulps there are semi-chemical pulps that use a mild chemical treatment followed by refining. Semi-chemical pulp is often used for corrugating medium.

Bales of recycled paper (normally old corrugated containers) for unbleached (brown) packaging grades may be simply pulped, screened and cleaned. Recycling to make white papers is usually done in a deinking plant, which employs screening, cleaning, washing, bleaching and flotation. Deinked pulp is used in printing and writing papers and in tissue, napkins and paper towels. It is often blended with virgin pulp.

At integrated pulp and paper mills, pulp is usually stored in high density towers before being pumped to stock preparation. Non integrated mills use either dry pulp or wet lap (pressed) pulp, usually received in bales. The pulp bales are slushed in a [re]pulper.

STOCK (PULP) PREPARATION

Stock preparation is the area where pulp is usually refined, blended to the appropriate proportion of hardwood, softwood or recycled fibre, and diluted to as uniform and constant as possible consistency. The pH is controlled and various fillers, such as whitening agents, size and wet strength or dry strength are added if necessary. Additional fillers such asclay, calcium carbonate and titanium dioxide increase opacity so printing on reverse side of a sheet will not distract from content on the obverse side of the sheet. Fillers also improve printing quality.

Pulp is pumped through a sequence of tanks that are commonly called *chests*, which may be either round or more commonly rectangular. Historically these were made of special ceramic tile faced reinforced concrete, but mild and stainless steels are also used. Low consistency pulp slurries are kept agitated in these chests by propeller like agitators near the pump suction at the chest bottom.

In the following process, different types of pulp, if used, are normally treated in separate but similar process lines until combined at a blend chest:

From high density storage or from slusher/pulper the pulp is pumped to a low density storage chest (tank). From there it is typically diluted to about 4% consistency before being pumped to an unrefined stock chest. From the unrefined stock chest stock is again pumped, with consistency control, through a refiner. Refining is an operation whereby the pulp slurry passes between a pair of discs, one of which is stationary and the other rotating at speeds of typically 1,000 or 1,200 RPM for 50 and 60 Hz AC, respectively. The discs have raised bars on their faces and pass each other with narrow clearance. This action unravels the outer layer of the fibres, causing the fibrils of the fibres to partially detach and bloom outward, increasing the surface area to promoting bonding. Refining thus increases tensile strength. For example, tissue paper is relatively unrefined whereas packaging paper is more highly refined. Refined stock from the refiner then goes to a refined stock chest, or blend chest, if used as such.

Hardwood fibres are typically 1 mm long and smaller in diameter than the 4 mm length typical of softwood fibres. Refining can cause the softwood fibre tube to collapse resulting in undesirable properties in the sheet. From the refined stock, or blend chest, stock is again consistency controlled as it is being pumped to a machine chest. It may be refined or additives may be added en route to the machine chest.

The machine chest is basically a consistency levelling chest having about 15 minutes retention. This is enough retention time to allow any variations in consistency entering the chest to be levelled out by the action of the basis weight valve receiving feedback from the on line basis weight measuring scanner. (Note: Many paper machines mistakenly control consistency coming out of the machine chest, interfering with basis weight control.)

SECTIONS OF THE PAPER MACHINE

There are four main sections on this paper machine. In practice calender rolls are normally placed vertically in a *stack*.

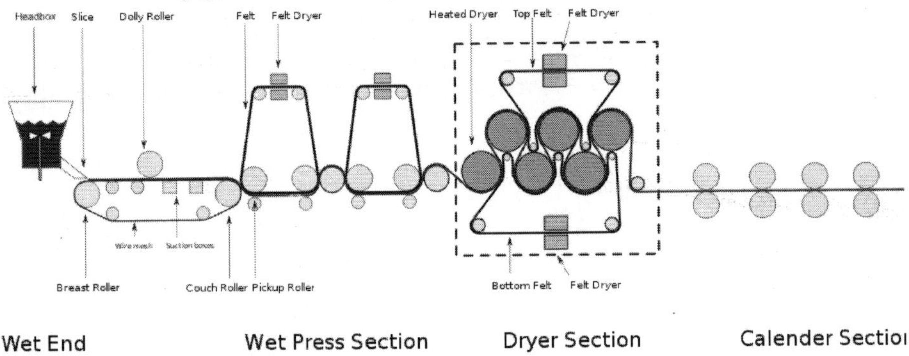

Fig. Diagram showing the sections of the Fourdrinier machine

Forming section or wet end

From the machine chest stock is pumped to a head tank, commonly called a *head box*, whose purpose is to maintain a constant head (pressure) on the stock as it feeds the basis weight valve. The head box also provides a means allowing air bubbles to escape. The basis weight valve meters the stock to the recirculating stream of water that is pumped, by the *fan pump*, from a whitewater chest through to the flow box. On the way to the flow box the pulp slurry may pass through centrifugal cleaners, which remove heavy contaminants like sand, and screens, which break up fibre clumps and remove over-sized debris.

Wood fibers have a tendency to attract one another, forming clumps, the effect being called flocculation. Flocculation is lessened by lowering consistency and or by agitating the slurry; however, de-flocculation becomes very difficult at much above 0.5% consistency. Minimizing the degree of flocculation when forming is important to physical properties of paper.

The consistency in the flow box is typically under 0.4% for most paper grades, with longer fibres requiring lower consistency than short fibres. Higher consistency causes more fibres to be oriented in the z direction, while lower consistency promotes fibre orientation in the x-y direction. Higher consistency promotes higher calliper (thickness) and stiffness, lower consistency promotes higher tensile and some other strength properties and also improves formation (uniformity). Many sheet properties continue to improve down to below 0.1% consistency; however, this is an impractical amount of water to handle. (Most paper machine run a higher headbox consistency than optimum because they have been sped up over time without replacing the fan pump and headbox. There is also an economic trade off with high pumping costs for lower consistency).

The stock slurry, often called *white water* at this point, exits the flow box through a rectangular opening of adjustable height called the *slice*, the white water stream being called the *jet* and it is pressurized on high speed machines so as to land gently on the moving fabric loop or *wire* at a speed typically between plus or minus 3% of the wire speed, called *rush* and *drag* respectively. Excessive *rush* or *drag* causes more orientation of fibres in the machine direction and gives differing physical properties in machine and cross directions; however, this phenomenon is not completely avoidable on Fourdrinier machines.

On lower speed machines at 700 feet per minute, gravity and the height of the stock in the headbox creates sufficient pressure to form the jet through the opening of the slice. The height of the stock is the head, which gives the headbox its name.

The speed of the jet compared to the speed of the wire is known as the *jet-to-wire ratio*. When the jet-to-wire ratio is less than unity, the fibres in the stock become drawn out in the machine direction. On slower machines where sufficient liquid remains in the stock before draining out, the wire can be driven back and forth with a process known as *shake*. This provides some measure of randomizing the direction of the fibres and gives the sheet more uniform strength in both the machine and cross-machine directions. On fast machines, the stock does not remain on the wire in liquid form long enough and the long fibres line up with the machine. When the jet-to-wire ratio exceeds unity, the fibers tend to pile up in lumps. The resulting variation in paper density provides the antique or parchment paper look.

Two large rolls typically form the ends of the drainage section, which is called the *drainage table*. The *breast roll* is located under the flow box, the jet being aimed to land on it at about the top centre. At the other end of the drainage table is the suction (*couch*) roll. The couch roll is a hollow shell, drilled with many thousands of precisely spaced holes of about 4 to 5 mm diameter. The hollow shell roll rotates over a stationary suction box, normally placed at the top centre or rotated just down machine. Vacuum is pulled on the suction box, which draws water from the web into the suction box. From the suction roll the sheet feeds into the press section.

Down machine from the suction roll, and at a lower elevation, is the *wire turning roll*. This roll is driven and pulls the wire around the loop. The wire turning roll has a considerable angle of wrap in order to grip the wire.

Supporting the wire in the drainage table area are a number of drainage elements. In addition to supporting the wire and promoting drainage, the elements de-flocculate the sheet. On low speed machines these table elements are primarily *table rolls*. As speed increases the suction developed in the nip of a table roll increases and at high enough speed the wire snaps back after leaving the vacuum area and causes stock to jump off the wire, disrupting the formation. To prevent this drainage foils are used. The foils are typically sloped between zero and two or three degrees and give a more gentle action. Where rolls and foils are used, rolls are used near the headbox and foils further down machine.

Approaching the dry line on the table are located low vacuum boxes that are drained by a barometric leg under gravity pressure. After the dry line are the suction boxes with applied vacuum. Suction boxes extend up to the couch roll. At the couch the sheet consistency should be about 25%.

Variations of the Fourdrinier forming section

The forming section type is usually based on the grade of paper or paperboard being produced; however, many older machines use a less than optimum design. Older machines can be upgraded to include more appropriate forming sections.

A second headbox may be added to a conventional fourdrinier to put a different fibre blend on top of a base layer. A *secondary headbox* is normally located at a point where the base sheet is completely drained. This is not considered a separate ply because the water action does a good job of intermixing the fibers of the top and bottom layer. Secondary headboxes are common on linerboard.

A modification to the basic fourdrinier table by adding a second wire on top of the drainage table is known as a top wire former. The bottom and top wires converge and some drainage is up through the top wire. A top wire improves formation and also gives more drainage, which is useful for machines that have been sped up.

The Twin Wire Machine or Gap former uses two vertical wires in the forming section, thereby increasing the de-watering rate of the fibre slurry while also giving uniform two sidedness.

There are also machines with entire Fourdrinier sections mounted above a traditional Fourdrinier. This allows making multi-layer paper with special characteristics. These are called top Fourdriniers and they make multi-ply paper or paperboard. Commonly this is used for making a top layer of bleached fibre to go over an unbleached layer.

Another type forming section is the cylinder mould machine using a mesh-covered rotating cylinder partially immersed in a tank of fibre slurry in the wet end to form a paper web, giving a more random distribution of the cellulose fibres. Cylinder machines can form a sheet at higher consistency, which gives a more three dimensional fibre orientation than lower consistencies, resulting in higher calliper (thickness) and more stiffness in the machine direction (MD). High MD stiffness is useful in food packaging like cereal boxes and other boxes like dry laundry detergent.

Tissue machines typically form the paper web between a wire and a special fabric (felt) as they wrap around a forming roll. The web is pressed from the felt directly onto a large diameter dryer called a *yankee*. The paper sticks to the yankee dryer and is peeled off with a scraping blade called a *doctor*. Tissue machines operate at speeds of up to 2000 m/min.

Press section

Fig. Paper machine

The second section of the paper machine is the press section, which removes much of the remaining water via a system of nips formed by rolls pressing against each other aided by press felts that support the sheet and absorb the pressed water. The paper web consistency leaving the press section can be above 40%.

Pressing is the most efficient method of de-watering the sheet as only mechanical action is required. Press felts historically were made from wool. However, today they are nearly 100% synthetic. They are made up of a polyamide woven fabric with thick batt applied in a specific design to maximise water absorption.

Presses can be single or double felted. A single felted press has a felt on one side and a smooth roll on the other. A double felted press has both sides of the sheet in contact with a press felt. Single felted nips are useful when mated against a smooth roll (usually in the top position), which adds a two-sidedness—making the top side appear smoother than the bottom. Double felted nips impart roughness on both sides of the sheet. Double felted presses are desirable for the first press section of heavy paperboard.

Simple press rolls can be rolls with grooved or blind drilled surface. More advanced press rolls are suction rolls. These are rolls with perforated shell and cover. The shell made of metal material such as bronze stainless steel is covered with rubber or a synthetic material. Both shell and cover are drilled throughout the surface. A stationary suction box is fitted in the core of the suction roll to support the shell being pressed. End face mechanical seals are used for the interface between the inside surface of the shell and the suction box. For the smooth rolls, they are typically made of granite rolls. The granite rolls can be up to 30-foot (9.1 m) long and 6 feet (1.8 m) in diameter.

Conventional roll presses are configured with one of the press rolls is in a fixed position, with a mating roll being loaded against this fixed roll. The felts run through the nips of the press rolls and continues around a felt run, normally consisting of several felt rolls. During the dwell time in the nip, the moisture from the sheet is transferred to the press felt. When the press felt exits the nip and continues around, a vacuum box known as an Uhle Box applies vacuum (normally -60 kPa) to the press felt to remove the moisture so that when the felt returns to the nip on the next cycle, it does not add moisture to the sheet.

Some grades of paper use suction pick up rolls that use vacuum to transfer the sheet from the couch to a lead in felt on the first press or between press sections. Pickup roll presses normally have a vacuum box that has two vacuum zones (low vacuum and high vacuum). These rolls have a large number of drilled holes in the cover to allow the vacuum to pass from the stationary vacuum box through the rotating roll covering.

The low vacuum zone picks up the sheet and transfers, while the high vacuum zone attempts to remove moisture. Unfortunately, at high enough speed centrifugal force flings out vacuumed water, making this less effective for dewatering. Pickup presses also have standard felt runs with Uhle boxes. However, pickup press design is quite different, as air movement is important for the pickup and dewatering facets of its role.

Crown Controlled Rolls (also known as CC Rolls) are usually the mating roll in a press arrangement. They have hydraulic cylinders in the press rolls that ensure that the roll does not bow. The cylinders connect to a shoe or multiple shoes to keep the crown on the roll flat, to counteract the natural "bend" in the roll shape due to applying load to the edges.

Extended Nip Presses (or ENP) are a relatively modern alternative to conventional roll presses. The top roll is usually a standard roll, while the bottom roll is actually a large CC roll with an extended shoe curved to the shape of the top roll, surrounded by a rotating rubber belt rather than a standard roll cover. The goal of the ENP is to extend the dwell time of the sheet between the two rolls thereby maximising the de-watering. Compared to a standard roll press that achieves up to 35% solids after pressing, an ENP brings this up to 45%

and higher—delivering significant steam savings or speed increases. ENPs densify the sheet, thus increasing tensile strength and some other physical properties.

Dryer section

Fig. Dryer section of an older Fourdrinier-style paper-making machine. These narrow, small diameter dryers are not enclosed by a hood, dating the photo to before the 1970s.

The dryer section of the paper machine, as its name suggests, dries the paper by way of a series of internally steam-heated cylinders that evaporate the moisture. Steam pressures may range up to 160 psig. Steam enters the end of the dryer head (cylinder cap) through a steam joint and condensate exits through a siphon that goes from the internal shell to a centre pipe. From the centre pipe the condensate exits through a joint on the dryer head. Wide machines require multiple siphons. In fast machines centrifugal force holds the condensate layer still against the shell and turbulence generating bars are typically used to agitate the condensate layer and improve heat transfer.

The sheet is usually held against the dryers by long felt loops on the top and bottom of each dryer section. The felts greatly improve heat transfer. Dryer felts are made of coarse thread and have a very open weave that is almost see through, It is common to have the first bottom dryer section unfelted to dump broke on the basement floor during sheet breaks or when threading the sheet.

Paper dryers are typically arranged in groups called *sections* so that they can be run at a progressively slightly slower speed to compensate for sheet shrinkage as the paper dries. The gaps between sections are called *draws*.

The drying sections are usually enclosed to conserve heat. Heated air is usually supplied to the pockets where the sheet breaks contact with the driers. This increases the rate of drying. The pocket ventilating tubes have slots along their entire length that face into the pocket. The dryer hoods are usually exhausted with a series of roof mounted hood exhausts fans down the dryer section.

SC Sizer

Additional sizing agents, including resins, glue, or starch, can be added to the web to alter its characteristics. Sizing improves the paper's water resistance, decreases its ability to fuzz, reduces abrasiveness, and improves its printing properties and surface bond strength. These may be applied at the wet (internal sizing) or on the dry end (surface sizing), or both. At the dry end sizing is usually applied with a *size press*. The size press may be a roll applicator (flooded nip) or Nozzle applicator . It is usually placed before the last dryer section. Some paper machines also make use of a 'coater' to apply a coating of fillers such as calcium carbonate or china clay usually suspended in a binder of cooked starch and styrene-butadiene latex. Coating produces a very smooth, bright surface with the highest printing qualities.

Calender section

A calender consists of two or more rolls, where pressure is applied to the passing paper. Calenders are used to make the paper surface extra smooth and glossy. It also gives it a more uniform thickness. The pressure applied to the web by the rollers determines the finish of the paper.

After calendering, the web has a moisture content of about 6% (depending on the furnish). It is wound onto a roll called a*tambour* or *reel*, and stored for final cutting and shipping. The roll hardness should be checked, obtained and adjusted accordingly to insure that the roll hardness is within the acceptable range for the product.

BANANA PAPER

Fig. Banana plant paper.

Banana paper is used in two different senses: the one refers to a paper made from the bark of the banana plant,and which is mainly used for artistic purposes; the other to paper made from banana fiber obtained, through an industrial process, from stems and the non-utilizable fruits. This paper can be either hand or machine made.

The banana agricultural industry processes 42 million tons of bananas every year (with 20,000 square kilometres planted). As a result of pulling apart the banana bunches from the main stem, there are leftover stems which contain 5% of fiber useful for the manufacture of paper.

BOND PAPER

Bond paper is a high quality durable writing paper similar to bank paper but having a weight greater than 50 g/m^2. The name comes from it having originally been made for documents such as government bonds. It is now used for letterheads, other stationery and as paper for electronic printers. Widely employed for graphic work involving pencil, pen and felt-tip marker, bond paper can sometimes contain rag fibre pulp, which produces a stronger, though rougher, sheet of paper.

BOOK PAPER

A book paper (or publishing paper) is a paper that is designed specifically for the publication of printed books.

Traditionally, book papers are off-white or low-white papers (easier to read), are opaque to minimise the show-through of text from one side of the page to the other, and are (usually) made to tighter caliper or thickness specifications, particularly for case-bound books. Typically, books papers are light-weight papers 60 to 90 g/m^2 and often specified by their caliper/substance ratios (volume basis). For example, a bulky 80 g/m^2 paper may have a caliper of 120 micrometres (0.12 mm) which would be Volume 15 ($120 \times 10/80$), whereas a low bulk 80 g/m^2 may have a caliper of 88 micrometres, giving a volume 11. This volume basis then allows the calculation of a book's PPI (printed pages per inch), which is an important factor for the design of book jackets and the binding of the finished book.

Different paper qualities can be used as book paper depending on the type of book. Machine-finished coated papers, woodfree uncoated papers, coated fine papers, andspecial fine papers are common paper grades.

COATED PAPER

Coated paper is paper which has been coated by a compound or polymer to impart certain qualities to the paper, including weight, surface gloss, smoothness or reduced ink absorbency. Kaolinite, calcium carbonate, Bentonite, talc... are used to coat paper for high quality printing used in packaging industry and in magazines. The chalk or china clay is bound to the paper with synthetic viscofiers, such as styrene-butadiene latexes and natural organic binders such as starch. The coating formulation may also contain chemical additives as dispersants, resins, PE: to give water resistance and wet strength to the paper, or to protect against ultraviolet radiation.

Varieties

Machine-finished coated paper

Machine-finished coated paper (MFC) has a basis weight of 48–80 g/m^2. They have good surface properties, high print gloss and adequate sheet stiffness. MFC papers are made of 60–85% groundwood or TMP and 15–40% chemical pulp with a total pigment content of 20–30%. The paper can be soft nip calendered or supercalendered. These are often used in paperbacks.

Coated fine paper

Coated fine paper or *woodfree coated paper* (WFC) are mainly produced for offset printing:

Standard coated fine papers

This paper quality is normally used for advertising materials, books, annual reports and high quality catalogs. Grammage ranges from 90–170 g/m^2 and ISO brightnessbetween 80–96%. The fibre furnish consists of more than 90% chemical pulp. Total pigment content are in the range 30–45%, where calcium carbonate and clay are the most common.

Low coat weight papers

These paper grades have lower coat weights than the standard WFC (3–14 g/m^2/side) and the grammage and pigment content are also generally lower, 55–135 g/m^2 and 20–35% respectively.

Art papers

Art papers are one of the highest quality printing papers and are used for illustrated books, calendars and brochures. The grammage varies from 100 to 230 g/m^2. These papergrades are triple coated with 20 - > 40 g/m^2/side and have matte or glossy finish. Higher qualities often contain cotton.

Others

Other types of paper coatings include polyethylene or polyolefin extrusion coating, silicone, and wax coating to make release liners, paper cups and photographic paper.Biopolymer coatings are available as more sustainable alternatives to common petrochemical coatings.

COTTON PAPER

Cotton paper is made from cotton linters or cotton from used cloths (rags) as the primary material source, hence the name rag paper. Cotton paper is superior in both strength and durability to wood pulp-based paper, which may contain high concentrations of acids.

PROPERTIES

Certain cotton fibre paper is known to last hundreds of years without appreciable fading, discoloration, or deterioration, so it is often used for important documents such as the archival copies of dissertations or theses. As a rule of thumb, for each percentage point of cotton fibre, a user may expect one year of resisting deterioration by use (the handling to which paper may be subjected). Legal document paper typically contains 25% cotton. Cotton paper will produce a better printout than copy paper because it is able to absorb ink/toner better.

Cotton paper is typically graded as 25%, 50%, or 100% cotton. Usually it can be checked by holding the cotton paper up to the light and looking just below the watermark for a number. 100% Cotton paper may contain small amounts of acids, and should be tested or certified before use for archival documents.

Second-cut cotton linters have a normal average fibre length of 1.45 μm, and have similar properties as a short softwood pulp.

Uses

Cotton paper is used in some countries' modern banknotes. These banknotes are typically made from 100% cotton paper, but can also be made from a mixture of 75% or lessflax. Other materials may also be used and still be known as *Currency paper.*

Cotton bond paper can be found at most stores that sell stationery and other office products. Some cotton paper contains a watermark.

Higher quality art papers are often made from cotton.

Also it has found extensive use as a Printed Circuit Board substrate when mixed with epoxy resins and classified into CEM 1, CEM 2 etc.

HISTORY

Cotton was first used with a mixture of silk to make paper called Carta Bombycina. In the 1800s, fiber crops such as flax fibres or cotton from used cloths (rags) were the primary material source. By the turn of the 20th century, most paper was made from wood pulp, but cotton is still used in specialty papers. As cotton rags now often contain synthetic fibres, papermakers have turned to second-cut cotton linters as raw material sources for making pulp for cotton papers.

INKJET PAPER

Fig. Paper for inkjet and laser printers in a store.

Inkjet paper is a special fine paper designed for inkjet printers, typically classified by its weight, brightness and smoothness, and sometimes by its opacity.

MANUFACTURE

Some inkjet papers are made from high quality deinked pulp or chemical pulps. Quality inkjet paper requires good dimensional stability, no curling or cockling, good surface strength. For most purposes surface smoothness is required. Sufficient and even porosity is required to counteract spreading of the ink. For lower quality printing, uncoated copy paper will suffice, but higher grades require coating. The traditional coatings are not widely used for inkjet papers. For matte inkjet papers, it is common to use silica as pigment together with polyvinyl alcohol (PVOH). Glossy inkjet papers can be made by multicoating, resin coating, or cast coating on a lamination paper.

There also exist a wide variety of fine art papers which meet the needs of professional photographers and artists. These papers share many characteristics with traditional watercolor, printmaking, and photographic papers. Fine art inkjet papers are designed to meet similar standards for longevity as traditional fine art papers: they should have a neutral pH, be lignin-free, and not include optical brighteners.

Fine art inkjet papers differ from traditional fine art papers, in that they include coatings which are designed to receive and hold inks. Fine art papers are usually made of rag pulp (100% cotton being the most common) but may also be have an alpha-cellulose base. Some descriptions and comparisons of fine art inkjet papers are here, here, here, and here. Some fine art papers are mold made, while others are machine made, and may vary considerably in surface texture. Many fine art papers are available in pre-cut sheets or in rolls.

COMPARISON TO STANDARD OFFICE PAPER

Standard office paper has traditionally been designed for use with typewriters and copy machines, where the paper usually does not get wet. With these types of paper, moisture tends to wick through the fibers away from the point of contact to form a disk. For an inkjet paper, this spreading results in the ink spreading out in the fibers to form a large smudge which lacks pigment intensity.

High-quality inkjet printing with dark, crisp lines requires the paper to have exactly the right degree of absorbency to accept the ink but prevent its sideways spread. Many general-purpose office papers of weights around 21 to 27 lb (80–100 g/m^2) have been reformulated so that they can be used equally well with both inkjet and laser printers. However, this category of paper is only suitable for printing text, because the ink load is light.

When paper is manufactured, it is formed from a fiber mat that collects on an open mesh screen, which is then dried and pressed flat and smooth. Large areas of inkjet color, such as found in graphics and photographs, soak the paper fibers with so much moisture that they swell and return to their original shape from before pressing, resulting in a wavy buckling of the paper surface.

Double-sided inkjet printing is usually not possible with inexpensive low-weight copy paper because of bleed-through from one side to the other. Heavier weight paper works better due to the thickness of the fibers limiting bleed-through.

These papers are also unsuitable for photographic work because standard office paper is usually not "white" enough. This results in a poor color gamut and leads to colors being described as "muddy".

For all types of paper, the settings in the printer driver must be adjusted to suit the paper, so that the right amount of ink is delivered.

INKJET PHOTO PAPER

Photo paper is a category of inkjet paper designed specifically for reproduction of photographs, which is extremely bright white due to bleaching or substances such as titanium dioxide, and has been coated with a highly absorbent material that limits diffusion of the ink away from the point of contact. Highly refined clay is a common coating to prevent ink spread.

The best of these papers, with suitable pigment-based ink systems, can match or exceed the image quality and longevity of photographic gelatin-based silver halide continuous tone printing methods used for color photographs, such as Fuji CrystalArchive (for color prints from negatives) and Cibachrome/ Ilfochrome (for color prints from positive transparencies). For printing monochrome photographs, traditional silver-based papers are widely felt to retain some advantage over inkjet prints. Photo paper is usually divided into glossy, semi-matte, semi-gloss, "satin" or "silk", and matte finishes. The thickness of photo paper varies over a wide range. The lighter weights are not much different from general-purpose office papers as described above, and can be used for all types of printing, although these are the least expensive lowest-quality photo paper.

Photo papers for more critical work are thicker and have advanced coatings, sometimes with quick-drying properties. They can normally be printed only on the one specially coated side. A few papers are coated for double-sided printing.

Glossy photo paper, which is generally the most popular, has a shiny finish that gives photos a vivid look. It will generally be smooth to the touch and will have some glare to it. Matte photo paper is less shiny and has less of a glare than glossy paper.

It is often used to produce superior text results. Matte and glossy prints will typically feel different to the touch, but when displayed under glass their results will often look quite similar.

To increase the resemblance to oil paintings, papers with an imitation canvas texture are available. Photo papers are usually high-brightness neutral white papers, but a few off-white papers are made.

As in offset litho printing and traditional photographic printing, glossy papers give the highest color density (or D_{max}), and therefore the widest color gamut. Photo papers vary in their longevity and their color gamut. Ink suppliers often provide color profiles for their ink systems when used with specific papers. Longevity depends on the specific combination of inks and paper. For maximum life, the paper substrate will be "woodfree" (i.e. wood-based but without lignin), or cotton-based, or a combination of the two. Plastic substrates also exist.

No official paper industry definition exists for glossy, semi-matte, etcetera, although an objective scale is available for the glossiness of papers used in offset litho printing. Such paper suppliers as Hewlett-Packard, Epson and Kodak all use their own terms to describe their paper, such as Everyday (HP), Premium High Gloss and Luster (Epson) and Ultima (Kodak). ECI (www.eci.org) has categorized papers for proofing simulation of litho papers (type1/2 etc.)

TYPES OF PAPER

Glossy

A type of paper that shines by nature. When light falls on such paper it is either reflected parallelly or at an angle. One has to be very careful while handling a glossy paper as it may get spots of fingers on it.

Luster

Luster papers are like glossy papers but they do not have a reflection problem that glossy paper has when a glossy paper photo is seen at an angle.

Metallic

With metallic paper a sheet of mylar is kept between the printing paper and the emulsion.

KRAFT PAPER

Fig. A piece of kraft paper

Kraft paper or kraft is paper or paperboard (cardboard) produced from chemical pulp produced in the kraft process.

Sack kraft paper, or just *sack paper*, is a porous kraft paper with high elasticity and high tear resistance, designed for packaging products with high demands for strength and durability.

Pulp produced by the kraft process is stronger than that made by other pulping processes; acidic sulfite processes degrade cellulosemore, leading to weaker fibers, and mechanical pulping processes leave most of the lignin with the fibers, whereas kraft pulping removes most of the lignin present originally in the wood. Low lignin is important to the resulting strength of the paper, as the hydrophobic nature of lignin interferes with the formation of the hydrogen bonds between cellulose (and hemicellulose) in the fibers.

Kraft pulp is darker than other wood pulps, but it can be bleached to make very white pulp. Fully bleached kraft pulp is used to make high quality paper where strength, whiteness and resistance to yellowing are important.

MANUFACTURE

Wood pulp for sack paper is made from softwood by the kraft process. The long fibers provide the paper its strength and wet strength chemicals are added to even further improve the strength. Both white and brown grades are made. Sack paper is then produced on a paper machine from the wood pulp. The paper is microcrepped to give porosity and elasticity. Microcrepping is done by drying with loose draws allowing it to shrink. This causes the paper to elongate 4% in the machine direction and 10% in the cross direction without breaking. Machine direction elongation can be further improved by pressing between very elastic cylinders causing more microcrepping. The paper may be coated with polyethylene (PE) to ensure an effective barrier against moisture, grease and bacteria. A paper sack can be made of several layers of sack paper depending on the toughness needed.

Kraft paper is produced on paper machines with moderate machine speeds. The raw material is normally softwood pulp from the kraft process.

Maintaining a high effective sulfur ratio or sulfidity is important for the highest possible strength using the kraft process.

The kraft process can use a wider range of fiber sources than most other pulping processes. All types of wood, including very resinous types like southern pine, and non-wood species like bamboo and kenaf can be used in the kraft process.

Qualities

- Normal kraft paper is strong and relatively coarse. It has high tensile strength. The grammage is normally from 40–135 g/m^2.
- Sack kraft paper, or just sack paper, is a porous kraft paper with high elasticity and high tear resistance, designed for packaging products with high demands for strength and durability.
- Absorbent kraft paper is made with controlled absorbency, i.e., a high degree of porosity. It is made of clean low kappa hardwood kraft and has to have a good uniformity and formation.
- Spinning kraft paper is an especially strong type of kraft paper with relatively low grammage (40 g/m^2). This paper requires the best possible machine direction strength and cross machine elongation. This is done by high fiber orientation on the papermachine.
- Hunting cartridge paper is a kraft paper used in shotgun shells. This paper needs a high tensile strength in the machine direction, which is the axial direction of the cartridges. In the cross direction, the cartridge is supported by the gun-pipe, but a sufficient elongation is needed. The body of the cartridge is wound of a kraft paper of 80–120 g/m^2, which is further covered by an outer sheet of 60–80 g/m^2 with colour and printing.
- Candy wrapping paper and twisting paper are thin 30–40 g/m^2 kraft papers and are mostly flexo or offset printed. These papers require a good strength, with highly oriented fibers. Twisting paper is mostly opaque and often supercalendered.

Applications

- As a plastic hazard free paper in paper sacks for cement, food, chemicals, consumer goods, flour bags etc.

- Electrical insulation in large oil-filled transformers.
- Paper grocery bags, multiwall sacks, envelopes and other packaging.
- An inexpensive material for lining particle boards.
- The base paper for sandpaper.

TYVEK

Tyvek is a brand of flashspun high-density polyethylene fibers, a synthetic material; the name is a registered trademark ofDuPont. It is often seen used as housewrap, a synthetic material used to protect buildings during construction. The material is very strong; it is difficult to tear but can easily be cut with scissors or a knife. Water vapor can pass through Tyvek, but liquid water cannot. All of these properties make Tyvek useful in a variety of applications.

HISTORY

Tyvek is a nonwoven product consisting of spunbond olefin fiber. It was first discovered in 1955 by DuPont researcher Jim White who saw polyethylene fluff coming out of a pipe in a DuPont experimental lab. It was trademarked in 1965 and was first introduced for commercial purposes in April 1967.

According to DuPont's website, the fibers are 0.5–10 μm (compared to 75 μm for a human hair). The nondirectional fibers (plexifilaments) are first spun and then bonded together by heat and pressure, without binders.

Tyvek is manufactured at the Spruance plant in Richmond, Virginia, and in Contern, Luxembourg.

Properties

Among Tyvek's properties are:

- Light weight
- Class A flammability rating.
- Chemical resistance
- Dimensional stability
- Opacity
- Neutral pH
- Tear resistance

ADHESION AND BONDING

Because it is essentially a form of plastic, some adhesives will not bond well with Tyvek. However, DuPont recommends starch, dextrin, casein, and animal-based adhesives over most synthetic-based adhesives, emphasizing the effectiveness of water-based and quick-drying glues as the best for bonding Tyvek both to itself and to a variety of substrates. DuPont also claims that the following adhesives are highly effective:

- Water-based synthetic lattices

- Ethylene/vinyl acetate
- Acrylic pressure-sensitive tape
- Solvent-based single-component polyurethane
- Hot glue

Heat-sealing can be used to melt the plastics in Tyvek and cause it to bond to itself, but this form of bonding tends to create puckers in the otherwise flat material. Dielectricbonding can be effective in some circumstances, as is ultrasonic sealing.

Uses

Large sheets of Tyvek are frequently used as "housewrap," to provide a water barrier between the outer cladding of a structure and the frame, insulation, etc., allowing water vapor to pass yet restricting air infiltration.

Tyvek is used by the United States Postal Service for some of its Priority Mail and Express Mail envelopes. New Zealand used it for itsdriver's licenses from 1986 to 1999, and Costa Rica, the Isle of Man, and Haiti have made banknotes from it. These banknotes are no longer in circulation and have become collectors' items.

Tyvek coveralls are one-piece garments, usually white, commonly worn by mechanics, painters, insulation installers, and laboratory andcleanroom workers where a disposable, one-time use coverall is needed. They are also used for some light HAZMAT applications, such as asbestos and radiation work but do not provide the protection of a full hazmat suit. Tychem is a sub-brand of Tyvek rated for a higher level of liquid protection, especially from chemicals. DuPont makes Tyvek clothing in different styles from laboratory coats and aprons to complete head-to-toe coveralls with hoods and booties. The latter was notably used by the Japan Ground Self-Defense Force as emergency limitedCBRN gear during the Fukushima nuclear incident.

In 1976, fashion house Fiorucci made an entire collection out of Tyvek. More recently fashion retailer and manufacturer American Apparel has included white Tyvek shorts as part of its range. Rock band Devo is known for wearing large, two-piece Tyvek suits with black elastic belts and 3-D glasses. In 1979, Devo appeared with Tyvek leisure suits and shirts made specifically for the band, with the band's own designs and images. In 2005, Dynomighty Design introduced a Tyvek wallet made from a single sheet of Tyvek. The ultralight backpackingcommunity has begun to use Tyvek for the construction of extremely light yet durable backpacks. In 2012, The Open Company released a foldable city map made of one of the stiffer variants of Tyvek.

Increasingly, reused Tyvek material is being used by home crafters. Protective sleeves for Compact Discs and DVDs, tote bags, and origami wallets also use Tyvek-containing materials.

Tyvek is also used as a durable fabric in shoes.

Tyvek is used in archery to construct waterproof target faces, replacing paper.

It is also extensively used for packaging in laboratory and medical equipment as the material withstands conditions used to sterilize equipment.

Tyvek wristbands are used at festivals and events where admission and security are concerns.

Wide Variety of Applications:

Pharmaceutical, electronics, diagnostic kits, foods, clothing, powders, vitamins, consumer products, parts, etc..

The Anna University, Chennai, TN, India uses Tyvek Paper for its degree certificates since 20th Dec 2010.

Recycling

Though Tyvek superficially resembles paper (for example, it can be written and printed on), it is plastic, and it cannot be recycled with paper. Some Tyvek products are marked with the #2 resin-code for HDPE, and can be collected with plastic bottles as part of some municipal curbside recycling programs. DuPont runs a program in the United States where disposable clothing, coveralls, lab coats, medical packaging and other non-hazardous Tyvek disposable garments can be recycled, as well as providing a mail-in recycling program for envelopes.

Recently, plastic bag recycling has become more prevalent. According to the American Chemistry Council, these plastic film drop-off locations accept Tyvek.

WALLPAPER

Wallpaper is a kind of material used to cover and decorate the interior walls of homes, offices, cafes, government buildings, museums, post offices, and other buildings; it is one aspect of interior decoration. It is usually sold in rolls and is put onto a wall usingwallpaper paste. Wallpapers can come plain as "lining paper" (so that it can be painted or used to help cover uneven surfaces and minor wall defects thus giving a better surface), textured (such as Anaglypta), with a regular repeating pattern design, or, much less commonly today, with a single non-repeating large design carried over a set of sheets.

Wallpaper printing techniques include surface printing, gravure printing, silk screen-printing, rotary printing, and digital printing. Wallpaper is made in long rolls, which are hung vertically on a wall. Patterned wallpapers are designed so that the pattern "repeats", and thus pieces cut from the same roll can be hung next to each other so as to continue the pattern without it being easy to see where the join between two pieces occurs. In the case of large complex patterns of images this is normally achieved by starting the second piece halfway into the length of the repeat, so that if the pattern going down the roll repeats after 24 inches the next piece sideways is cut from the roll to begin 12 inches down the pattern from the first. The number of times the pattern repeats horizontally across a roll does not matter for this purpose. A single pattern can be issued in several different colorways.

Fig. Hanging a traditional repeat pattern wallpaper

HISTORY

The main historical techniques are: hand-painting, woodblock printing (overall the most common), stencilling, and various types of machine-printing. The first three all date back to before 1700.

Wallpaper, using the printmaking technique of woodcut, gained popularity in Renaissance Europe amongst the emerging gentry. The social elite continued to hang largetapestries on the walls of their homes, as they had in the Middle Ages.

These tapestries added color to the room as well as providing an insulating layer between the stone walls and the room, thus retaining heat in the room. However, tapestries were extremely expensive and so only the very rich could afford them. Less well-off members of the elite, unable to buy tapestries due either to prices or wars preventing international trade, turned to wallpaper to brighten up their rooms.

Early wallpaper featured scenes similar to those depicted on tapestries, and large sheets of the paper were sometimes hung loose on the walls, in the style of tapestries, and sometimes pasted as today. Prints were very often pasted to walls, instead of being framed and hung, and the largest sizes of prints, which came in several sheets, were probably mainly intended to be pasted to walls. Some important artists made such pieces - notably Albrecht Dürer, who worked on both large picture prints and also ornament prints - intended for wall-hanging. The largest picture print was *The Triumphal Arch* commissioned by the Holy Roman Emperor Maximilian I and completed in 1515. This measured a colossal 3.57 by 2.95 metres, made up of 192 sheets, and was printed in a first edition of 700 copies, intended to be hung in palaces and, in particular, town halls, after hand-coloring.

Very few samples of the earliest repeating pattern wallpapers survive, but there are a large number of old master prints, often in engraving of repeating or repeatable decorative patterns. These are called ornament prints and were intended as models for wallpaper makers, among other uses.

England and France were leaders in European wallpaper manufacturing. Among the earliest known samples is one found on a wall from England and is printed on the back of a London proclamation of 1509. It became very popular in England following Henry VIII's excommunication from the Catholic Church - English aristocrats had always imported tapestries from Flanders and Arras, but Henry VIII's split with the Catholic Church had resulted in a fall in trade with Europe. Without any tapestry manufacturers in England, English gentry and aristocracy alike turned to wallpaper.

During the Protectorate under Oliver Cromwell, the manufacture of wallpaper, seen as a frivolous item by the Puritan government, was halted. Following the Restoration ofCharles II, wealthy people across England began demanding wallpaper again - Cromwell's regime had imposed a boring culture on people, and following his death, wealthy people began purchasing comfortable domestic items which had been banned under the Puritan state.

18th century

In 1712, during the reign of Queen Anne, a wallpaper tax was introduced which was not abolished until 1836. By the mid-eighteenth century, Britain was the leading wallpaper manufacturer in Europe, exporting vast quantities to Europe in addition to selling on the middle-class British market. However this trade was seriously disrupted in 1755 by the Seven Years' War and later the Napoleonic Wars, and by a heavy level of duty on imports to France.

In 1748 the British Ambassador to Paris decorated his salon with blue flock wallpaper, which then became very fashionable there. In the 1760s the French manufacturer Jean-Baptiste Réveillon hired designers working in silk and tapestry to produce some of the most subtle and luxurious wallpaper ever made. His sky blue wallpaper with fleurs-de-lys was used in 1783 on the first balloons by the Montgolfier brothers. The landscape painter Jean-Baptiste Pillement discovered in 1763 a method to use fast colours.

Hand-blocked wallpapers like these use hand-carved blocks and by the 18th century designs include panoramic views of antique architecture, exotic landscapes and pastoral subjects, as well as repeating patterns of stylized flowers, people and animals.

In 1785 Christophe-Philippe Oberkampf had invented the first machine for printing coloured tints on sheets of wallpaper. In 1799 Louis-Nicolas Robert patented a machine to produce continuous lengths of paper, the forerunner of the Fourdrinier machine. This ability to produce continuous lengths of wallpaper now offered the prospect of novel designs and nice tints being widely displayed in drawing rooms across Europe.

Wallpaper manufacturers active in England in the 18th century included John Baptist Jackson and John Sherringham. Among the firms established in 18th-century America: J. F. Bumstead & Co. (Boston), William Poyntell (Philadelphia), John Rugar (New York).

High-quality wallpaper made in China became available from the later part of the 17th century; this was entirely handpainted and very expensive. It can still be seen in rooms in palaces and grand houses including Nymphenburg Palace, £azienki Palace, Chatsworth House, Temple Newsam, Broughton Castle, Lissan House, and Erddig. It was made up to 1.2 metres wide. English, French and German manufacturers imitated it, usually beginning with a printed outline which was coloured in by hand, a technique sometimes also used in later Chinese papers.

France and America

Towards the end of the 18th century the fashion for scenic wallpaper revived in both England and France, leading to some enormous panoramas, like the 1804 20 strip wide panorama, *Sauvages de la Mer du Pacifique* (Savages of the Pacific), designed by the artist Jean-Gabriel Charvet for the French manufacturer Joseph Dufour et Cie showing the Voyages of Captain Cook. This famous so called "papier peint" wallpaper is still in situ in Ham House, Peabody Massachusetts.

It was the largest panoramic wallpaper of its time, and marked the burgeoning of a French industry in panoramic wallpapers. Dufour realized almost immediate success from the sale of these papers and enjoyed a lively trade with America. The Neoclassical style currently in favour worked well in houses of the Federal period with Charvet's elegant designs. Like most 18th-century wallpapers, the panorama was designed to be hung above a dado.

Beside Joseph Dufour et Cie (1797 - c. 1830) other French manufacturers of panoramic scenic and trompe l'œil wallpapers, Zuber et Cie (1797–present) and Arthur et Robertexported their product across Europe and North America. Zuber et Cie's c. 1834 design *Views of North America* hangs in the Diplomatic Reception Room of the White House.

While Joseph Dufour et Cie was shut down in the 1830s, Zuber et Cie still exists and with Cole & Son of England is one of the last Western producers of woodblock printed wallpapers.

For its production Zuber uses woodblocks out of an archive of more than 100,000 cut in the 19th century which are classified as a "Historical Monument". It offers panoramic sceneries such as "Vue de l'Amérique Nord", "Eldorado Hindoustan" or "Isola Bella" and also wallpapers, friezes and ceilings as well as hand-printed furnishing fabrics.

Among the firms begun in France in the 19th century: Desfossé & Karth. In the United States: John Bellrose, Blanchard & Curry, Howell Brothers, Longstreth & Sons, Isaac Pugh in Philadelphia; Bigelow, Hayden & Co. in Massachusetts; Christy & Constant, A. Harwood, R. Prince in New York.

England

During the Napoleonic Wars, trade between Europe and Britain evaporated, resulting in the gradual decline of the wallpaper industry in Britain. However, the end of the war saw a massive demand in Europe for British goods which had been inaccessible during the wars, including cheap, colourful wallpaper. The development of steam-powered printing presses in Britain in 1813 allowed manufacturers to mass-produce wallpaper, reducing its price and so making it affordable to working-class people.

Wallpaper enjoyed a huge boom in popularity in the nineteenth century, seen as a cheap and very effective way of brightening up cramped and dark rooms in working-class areas. It became almost the norm in most areas of middle-class homes, but remained relatively little used in public buildings and offices, with patterns generally being avoided in such locations. In the latter half of the century Lincrusta and Anaglypta, not strictly wallpapers, became popular competitors, especially below a dado rail. They could be painted and washed, and were a good deal tougher, though also more expensive.

Wallpaper manufacturing firms established in England in the 19th century included Jeffrey & Co.; Shand Kydd Ltd.; Lightbown, Aspinall & Co.; John Line & Sons; Potter & Co.; Arthur Sanderson & Sons; Townshend & Parker. Designers included Owen Jones, William Morris, and Charles Voysey. In particular, many 19th century designs by Morris and Co and other Arts and Crafts designers remain in production.

20th century

By the early twentieth century, wallpaper had established itself as one of the most popular household items across the Western world. Manufacturers in the USA included Sears; designers included Andy Warhol. Wallpaper has gone in and out of fashion since about 1930, but the overall trend has been for wallpaper-type patterned wallcoverings to lose ground to plain painted walls.

21st century

In the early XXI century wallpaper evolved into a lighting feature, enhancing the mood and the ambience through lights and crystals. Meystyle, London based company, invented LED incorporated wallpaper. The development of digital printing allows designers to break the mould and combine new technology and art to bring wallpaper to a new level of popularity.

Historical collections

Historical examples of wallpaper are preserved by cultural institutions such as the Deutsches Tapetenmuseum (Kassel) in Germany; the Musée des Arts Décoratifs (Paris) and Musée du Papier Peint (Rixheim) in France; the Victoria & Albert in the UK; the Smithsonian's Cooper-Hewitt, Historic New England, Metropolitan Museum of Art, U.S. National Park Service, and Winterthur in the USA.

TYPES AND SIZES

In terms of methods of creation, wallpaper types include painted wallpaper, hand-printed blockwood wallpaper, hand-printed stencil wallpaper, machine-printed wallpaper, and flock wallpaper.

Modern wallcoverings are diverse, and what is described as wallpaper may no longer actually be made from paper. Two of the most common factory trimmed sizes of wallpaper are referred to as "American" and "European" rolled goods. American rolled goods are 27 inches by 27 feet (8.2 m) in length. European rolled goods are 21.5 inches wide by 33 feet (10 m) in length. Approx. 60 square feet (5.6 m). Most wallpaper borders are sold by linear foot and with a wide range of widths therefore square footage is not applicable. Although some may require trimming.

The most common wall covering for residential use and generally the most economical is prepasted vinyl coated paper, commonly called "strippable" which can be misleading. Cloth backed vinyl is fairly common and durable. Lighter vinyls are easier to handle and hang. Paper backed vinyls are generally more expensive, significantly more difficult to hang, and can be found in wider untrimmed widths. Foil wallpaper generally has paper backing and can (exceptionally) be up to 36 inches wide, and be very difficult to handle and hang. Textile wallpapers include silks, linens, grass cloths, strings, rattan, and actual impressed leaves. There are acoustical wall carpets to reduce sound. Customized wallcoverings are available at high prices and most often have minimum roll orders.

Solid vinyl with a cloth backing is the most common commercial wallcovering and comes from the factory as untrimmed at 54 inches approximately, to be overlapped and double cut by the installer. This same type can be pre-trimmed at the factory to 27 inches approximately.

Furthermore, wallpaper comes in the form of borders, typically mounted horizontally, and commonly near ceiling level of homes. Borders come in varying widths and patterns.

MODERN DEVELOPMENTS

Custom wallpaper printing

New digital inkjet printing technologies using ultraviolet (UV) cured inks are being used for custom wallpaper production. Very small runs can be made, even a single wall. Photographs or digital art are output onto blank wallpaper material. Typical installations are corporate lobbies, restaurants, athletic facilities, and home interiors. This gives a designer the ability to give a space the exact look and feel desired.

High-tech wallpaper

New types of wallpaper under development or entering the market in the early 21st century include wallpaper that blocks certain mobile phone and WiFi signals, in the interest of privacy. The wallpaper is coated with a silver ink which forms crystals that block outgoing signals.

The Spanish firm Think Big Factory has announced that they are developing a wallpaper which also serves as a computer interface, using projectors, webcams, and motion sensors for control. As of 2013, The hardware was complete but only 20 per cent of the software was finished, according to Think Big Factory.

Seismic wallpaper

In 2012, Scientists at the Institute of Solid Construction and Construction Material Technology at the Karlsruhe Institute of Technology announced that they had developed a wallpaper that can help keep a masonry wall from failing in an earthquake. The wallpaper uses glass fibre reinforcement in several directions and a special adhesive which forms a strong bond with the masonry when dry.

Installation

Like paint, wallpaper requires proper surface preparation before application. Additionally wallpaper is not suitable for all areas. For example, bathroom wallpaper may deteriorate rapidly due to excessive steam. Proper preparation includes the repair of any defects in the drywall or plaster and the removal of loose material or old adhesives. For a better finish with thinner papers and poorer quality walls the wall can be cross-lined (horizontally) with lining paper first .Accurate room measurements (length, width, and height) along with number of window and door openings is essential for ordering wallpaper. Large drops, or repeats, in a pattern can be cut and hung more economically by working from alternating rolls of paper.

Besides conventional installation on interior walls and ceilings, wallpapers have been deployed as decorative covering for hatboxes, bandboxes, books, shelves, and window-shades.

Wallpaper adhesives

Most wallpaper adhesive are starch or methylcellulose based.

Removal

Water

The simplest removal option is to brush the paper with water. Water soaks through the paper and saturates the glue, allowing the paper to be peeled off.

This does not work well with non-peelable vinyls, as vinyl is not porous. Nevertheless, it is still effective on many modern papers.

A mixture of 3:1 or 1:1 water and white vinegar is effective at dissolving glues. If the wallpaper is scored or sanded with a 20 grit floor sanding pad to scratch the surface solution, uptake will be more effective.

Chemical wallpaper stripper

Chemical wallpaper stripper can be purchased at most paint or home improvement stores. It is mixed with warm water or a mixture of warm water and vinegar, then sprayed onto wall surfaces. Several applications may be required to saturate the existing wallpaper. Perforation can aid in the absorption of the mixture and lead to faster removal. After the mixture has dissolved the wallpaper paste, the wallpaper can be removed easily by pulling at the edges and with the aid of a putty or drywall knife.

Steam

Another method of removal is to apply steam to wallpaper in order to dissolve the wallpaper paste. A wallpaper steamer consists of a reservoir of water, an electric heating element, and a hose to direct the steam at the wallpaper. The steam dissolves the wallpaper paste, allowing the wallpaper to be peeled off. However, care must be taken to prevent damage to the drywall underneath. Sometimes steaming can lead to the crumbling of underlying drywall or plaster, leaving an uneven surface to be repaired.

WATERPROOF PAPER

Waterproof paper is good for extreme outdoor, marine, field use and general wet environments. Often designed especially for printing topographic maps. It is normally durable and tear-resistant.

The paper is created with special coatings and fibers to allow it to stay together and not change shape or texture when exposed to rain, dampness, or immersed in water.

Waterproof paper is difficult to recycle.

WAX PAPER

Wax paper or waxed paper, also known as paraffin paper, is paper that is made moisture-proof through the application of wax.

The practice of oiling parchment or paper in order to make it semi-translucent or moisture-proof goes back at least to the middle ages. Paper impregnated or coated with purified beeswax was widely used throughout the 19th century to retain or exclude moisture, or to wrap odorous products. Gustave Le Gray introduced the use of waxed paper for photographic negatives in 1851. Natural wax was largely replaced for the making of wax paper (or paraffine paper) after Herman Frasch developed ways of purifying paraffin and coating paper with it in 1876. Wax paper is commonly used in cooking, for its non-stick properties, and wrapping food for storage, such ascookies, as it keeps water out or in. It is also used in arts and crafts.

Food preparation

Oven: wax paper is not recommended for baking use as it will smoke.

Microwave: wax paper can be used to prevent splatters by covering the food when microwave cooking. Since the paper is mostly unaffected by microwaves, it will not heat to the point of combustion under normal usage. This makes wax paper more functional than plastic wrap which will melt at lower temperatures, or aluminium foil which is not safe for use in most microwave ovens.

Other uses

Safety razor blades are traditionally wrapped in wax paper to make handling them less dangerous. Wax paper can also be used to make long lasting paper boats because of its high resistance to water.

From the early 1950s to the mid-1990s, wax paper was used as a common wrapping for sports card packages (Oh-Pee-Chee, Topps, Donruss, etc.). It was notorious for leaving wax markings on the back card where the wax paper was heated to be sealed. Wax paper was used as a way to keep the enclosed piece of bubble gum protected.

In the mid-1990s, sports card manufacturers stopped including pieces of bubble gum in packs of sports cards, thus ending the need for wax paper packs. Plastic (mylar) or other plastic/paper blends were used from then on.

Wax paper is also commonly used to attach pattern pieces to fabric while cutting it for sewing. One presses an iron over the wax paper briefly and attaches it to the cloth, making it easier to trace while cutting. Wax paper's particularly high dielectric strength makes it a practical electrical insulator, although modern materials have surpassed and mostly replaced it. Common applications are coil winding separators and capacitor dielectrics, and other applications requiring resilience against a potential difference up to the order of a few thousand volts per layer.

Turntablists (DJ's) commonly place one or multiple sheets of wax paper under their records to increase record slip and aid in scratch (where the record is rotated in a number of different ways by a finger to create special sound effects) routines.

Environmental issues

There are multiple environmental issues concerned with wax paper. Though it is biodegradable in its unaltered form, oft-applied additives such as petroleum rid it of that quality . Wax paper also cannot be recycled.

WOVE PAPER

Wove paper is a writing paper with a uniform surface, not ribbed or watermarked. The papermaking mould's wires run parallel to each other to produce laid paper, but they are woven together into a fine wire mesh for wove paper. The originator of this new papermaking technique was James Whatman (1702–59) from Kent, England.

For 500 years European paper makers could only produce what came to be called laid paper. In 1757 John Baskerville printed his famous edition of Virgil on a new kind of paper, called Wove (known in Europe as Vélin). This paper is now known to have been made by the elder James Whatman. Twenty-five years later (1780s) the manufacture of wove paper spread quickly to other paper mills in England, and was also being developed in France and America. All this took place over a decade before a machine to replace making paper by hand was conceived. With the establishment of the paper machine (1807), the manufacture of paper on a wove wire base would become the predominant standard in the world, with laid paper being relegated to certain specialist uses, such as being used as a support for charcoal drawings. Today more than 99% of the world's paper is made in this way.

WHATMAN PAPER

Whatman paper is a type of wove paper named after James Whatman. It is notable for its exceptional quality. Whatman paper is grained, strong and rigid, without laid lines. It is used in publishing, filtering, and chromatography.

XUAN PAPER

Xuan paper or Shuen paper or rice paper, is a kind of paper originating in ancient China used for writing and painting. Xuan paper is renowned for being soft and fine textured, suitable for conveying the artistic expression of both Chinese calligraphy and painting.

ORIGIN

Xuan paper was first mentioned in ancient Chinese books *Notes of Past Famous Paintings* and *New Book of Tang*. It was originally produced in the Tang dynasty in Jing County, which was under the jurisdiction of Xuan Prefecture (Xuanzhou), hence the name Xuan paper. By the Song dynasty, the paper producing industries in Huizhou and Chizhou were gradually transferred to Jing County.

CLASSIFICATION

Due to different producing methods, Xuan paper can be classified into *Shengxuan*, *Shuxuan*, and *Banshuxuan*. Shengxuan (literally "Raw Xuan"), which is not specially processed, excels in its ability to absorb water, causing the ink on it to blur. Shuxuan (literally "Ripe Xuan"), however, has Potassium alum worked into it during production, which results in a stiffer texture, a reduced ability to absorb water, and less resistance to shear stress (meaning that it can be torn much more easily). This feature makes Shuxuan more suitable for Gongbi rather than Xieyi. Banshuxuan (literally "Half-ripe Xuan") has intermediate absorbability, between Shengxuan and Shuxuan.

FEATURES

Xuan paper features great tensile strength, smooth surface, pure and clean texture and clean stroke, great resistance to crease, corrosion, moth and mold. The majority of ancient Chinese books and paintings by famous painters that survived till today are well preserved on Xuan paper. Xuan paper won the Golden Award at the Panama International Exposition in 1915. Xuan paper was used to make scrolls.

MATERIAL AND PRODUCTION

The material Xuan paper uses is closely related to the geography of Jing County. The bark of *Pteroceltis tatarinowii*, a common species of elm in the area, was used as the main material to produce Xuan paper. Rice, along with several other materials, was subsequently added to the recipe in the Song and Yuan Dynasties. Bamboo and mulberry also began to be used to produce xuan paper around that time. The production of Xuan paper can be loosely described as an 18-step process, and a detailed account would involve over a hundred. Some paper makers have invented steps which have been kept secret from others. The process includes steaming and bleaching the bark of *Pteroceltis tatarinowii*, as well as the addition of a variety of juices.

ENVIRONMENTAL IMPACT OF PAPER

The environmental impact of paper is significant, which has led to changes in industry and behaviour at both business and personal levels. With the use of modern technology such as the printing press and the highly mechanised harvesting of wood, disposable paper has become a cheap commodity. This has led to a high level of consumption and waste. With the rise in environmental awareness due to the lobbying by environmental organizations and with increased government regulation there is now a trend towards sustainability in thepulp and paper industry.

ISSUES

The production and use of paper has a number of adverse effects on the environment which are known collectively as paper pollution. Pulp mills contribute to air, water andland pollution. Discarded paper is a major component of many landfill sites, accounting for about 35 percent by weight of municipal solid waste (before recycling). Even paper recycling can be a source of pollution due to the sludge produced during de-inking.

According to a Canadian citizens organization, “People need paper products and we need sustainable, environmentally safe production.” The amount of paper and paper products is enormous, so the environmental impact is also very significant. It has been estimated that by 2020 paper mills will produce almost 500,000,000 tons of paper and paperboard per year, so great efforts are needed to ensure that the environment is protected during the production, use and recycling/disposal of this enormous volume of material.

Pulp and paper is the third largest industrial polluter to air, water, and land in both Canada and the United States, and releases well over 100 million kg of toxic pollution each year.

Worldwide, the pulp and paper industry is the fifth largest consumer of energy, accounting for four percent of all the world’s energy use. The pulp and paper industry uses more water to produce a ton of product than any other industry.

Deforestation

Worldwide consumption of paper has risen by 400% in the past 40 years, with 35% of harvested trees being used for paper manufacture. Plantation forest, from where the majority of wood for pulping is obtained, is generally a monoculture and this raises concerns over the ecological effects of the practice.

In the United States, paper is often made of small trees thinned from forests at intervals of around fifteen years in the southern U.S. Thinning improves overall forest health in that it helps prevent infestations of pests such at beetles and decreases fire risk. Wood sold for pulp to make paper is worth significantly less than wood sold for lumber, so large trees are almost never used to make paper. Most paper mills are unable to take large trees.

Deforestation is often seen as a problem in developing countries but also occurs in the developed world. Woodchipping to produce paper pulp is a contentious environmental issue in Australia. In the 1990s, the New Zealand government stopped the export of woodchips from native forests after campaigning by environmentalists.

Over 6.5 million trees were cut down to make 16 billion paper cups used by US consumers only for coffee in 2006, using 4 billion US gallons (15,000,000 m^3) of water and resulting in 253 million pounds of waste. Overall, North Americans use 58% of all paper cups, amounting to a staggering 130 billion cups.

Air pollution

Nitrogen dioxide (NO_2) sulfur dioxide (SO_2) and carbon dioxide (CO_2) are all emitted during paper manufacturing. Nitrogen dioxide and sulfur dioxide are major contributors ofacid rain, whereas CO_2 is a greenhouse gas responsible for climate change.

Water pollution

Wastewater discharges for a pulp and paper mill contains solids, nutrients and dissolved organic matter such as lignin. It also contains alcohols, and chelating agents and inorganic materials like chlorates and transition metal compounds.

Nutrients such as nitrogen and phosphorus can cause or exacerbate eutrophication of fresh water bodies such as lakes and rivers. Organic matter dissolved in fresh water, measured by Biological Oxygen Demand (BOD), changes ecological characteristics, and in worst-case scenarios leads to death of all higher living organisms. Waste water may also be polluted with organochlorine compounds. Some of these are naturally occurring in the wood, but chlorine bleaching of the pulp produces far larger amounts.

Recycling the effluent and burning it, using bioremediation ponds and employing less damaging agents in the pulping and bleaching processes can help reduce water pollution.

Discharges can also discolour the water leading to reduced aesthetics. This has happened with the Tarawera River in New Zealand which subsequently became known as the "black drain".

Waste

Paper waste accounts for up to 40% of total waste in the United States, which adds up to 71.6 million tons of waste per year in the United States alone. Paper waste like other wastes faces the additional hazard of toxic inks, dyes and polymers that could be potentially carcinogenic when incinerated, or comingled with groundwater via traditional burial methods such as modern landfills. Paper recycling mitigates this impact, but not the environmental and economic impact of the energy consumed by manufacturing, transporting and burying and or reprocessing paper products.

Chlorine and chlorine-based materials

Chlorine and compounds of chlorine are used in the bleaching of wood pulp, especially chemical pulps produced by the kraft process or sulfite process. Plants using elemental chlorine produced significant quantities of dioxins. Dioxins are persistent organic pollutants that are generally recognized among the most toxic human-released pollutants in existence. Elemental chlorine has largely been replaced by chlorine dioxide and dioxin production very significantly reduced.

As a result, from the 1990 onwards the use of elemental chlorine in the delignification process was substantially reduced and replaced with ECF (Elemental Chlorine Free) and TCF (Totally Chlorine Free) bleaching processes. In 2005, elemental chlorine was used in 19–20% of kraft pulp production globally, down from over 90% in 1990. 75% of kraft pulp used ECF, with the remaining 5–6% using TCF. A study based on U.S. Environmental Protection Agency (EPA) data concluded that "Studies of effluents from mills that use oxygen delignification and extended delignification to produce ECF (elemental chlorine free) and TCF pulps suggest that the environmental effects of these processes are low and similar." Most TCF pulp is produced in Sweden and Finland for sale in Germany, all markets with a high level of environmental awareness. In 1999, TCF pulp represented 25% of the European market.

TCF bleaching, by removing chlorine from the process, reduces chlorinated organic compounds to background levels in pulp mill effluent. ECF bleaching can substantially reduce but not fully eliminate chlorinated organic compounds, including dioxins, from effluent. While modern ECF plants can achieve chlorinated organic compounds (AOX) emissions of less than 0.05 kg per tonne of pulp produced, most do not achieve this level of emissions. Within the EU, the average chlorinated organic compound emissions for ECF plants is 0.15 kg per tonne.

However, there has been disagreement about the comparative environmental effects of ECF and TCF bleaching. On the one hand, paper and chemical industry-funded studies have generally found that there is no environmental difference between ECF and TCF effluents. On the other hand, independent peer-reviewed study has found that, comparing conventional, ECF and TCF effluents before and after secondary treatment, "TCF effluents are the least toxic".

Sulfur, hydrogen sulfide, and sulfur dioxide

Sulfur-based compounds are used in both the kraft process and the sulfite process for making wood pulp. Sulfur is generally recovered, with the exception of ammonia-based sulfite processes, but some is released as sulfur dioxide during combustion of black liquor, a byproduct of the kraft process, or "red liquor" from the sulfite process. Sulfur dioxideis of particular concern because it is water-soluble and is a major cause of acid rain. In 2006 the pulp and paper industry in Canada released about 60,000 tonnes of sulfur oxides (SOx) into the atmosphere, accounting for just over 4% of the total SOx emission from all Canadian industries.

A modern kraft pulp mill is more than self-sufficient in its electrical generation and normally will provide a net flow of energy to the local electrical grid. Additionally, bark and wood residues are often burned in a separate power boiler to generate steam.

Air emissions of hydrogen sulfide, methyl mercaptan, dimethyl sulfide, dimethyl disulfide, and other volatile sulfur compounds are the cause of the odor characteristic of pulp mills utilizing the kraft process. Other chemicals that are released into the air and water from most paper mills include the following:

- Carbon monoxide
- Ammonia
- Nitrogen oxide
- Mercury
- Nitrates
- Methanol
- Benzene
- Volatile organic compounds, chloroform.

Bleaching mechanical pulp is not a major cause for environmental concern since most of the organic material is retained in the pulp, and the chemicals used (hydrogen peroxideand sodium dithionite) produce benign byproducts (water and, eventually, sodium sulfate, respectively).

However, the bleaching of chemical pulps has the potential to cause significant environmental damage, primarily through the release of organic materials into waterways. Pulp mills are almost always located near large bodies of water because they require substantial quantities of water for their processes. An increased public awareness of environmental issues from the 1970s and 1980s, as evidenced by the formation of organizations like Greenpeace, influenced the pulping industry and governments to address the release of these materials into the environment. Environmental NGO pressure was especially intense on Swedish and Finnish pulp and paper companies.

Conventional bleaching using elemental chlorine produces and releases into the environment large amounts of chlorinated organic compounds, including chlorinated dioxins.Dioxins are recognized as a persistent environmental pollutant, regulated internationally by the Stockholm Convention on Persistent Organic Pollutants.

Dioxins are highly toxic, and health effects on humans include reproductive, developmental, immune and hormonal problems. They are known to be carcinogenic. Over 90% of human exposure is through food, primarily meat, dairy, fish and shellfish, as dioxins accumulate in the food chain in the fatty tissue of animals.

Non-renewable resources

Clay or calcium carbonate are used as fillers for some papers. Kaolin is the most commonly used clay for coated papers.

MITIGATION

Some of the effect of the pulp and paper industry can be addressed and there is some change towards sustainable practices. The use of wood solely from plantation forests addresses concerns about loss of old growth forests.

Bleaching

The move to non-elemental chlorine for the bleaching process reduced the emission of the carcinogenic organochlorines. Peracetic acid,ozone and hydrogen peroxide and oxygen are used in bleaching sequences in the pulp industry to produce totally chlorine free (TCF) paper.

Recycling

There are three categories of paper that can be used as feedstocks for making recycled paper: mill broke, pre-consumer waste, andpost-consumer waste. Mill broke is paper trimmings and other paper scrap from the manufacture of paper, and is recycled internally in a paper mill. Pre-consumer waste is material that was discarded before it was ready for consumer use. Post-consumer waste is material discarded after consumer use such as old magazines, old telephone directories, and residential mixed paper.

One concern about recycling wood pulp paper is that the fibers are degraded with each and after being recycled four or five times the fibers become too short and weak to be useful in making paper.

The United States Environmental Protection Agency has found that recycling causes 35% less water pollution and 74% less air pollution than making virgin paper. Pulp millscan be sources of both air and water pollution, especially if they are producing bleached pulp. Recycling paper decreases the demand for virgin pulp and thus reduces the overall amount of air and water pollution associated with paper manufacture. Recycled pulp can be bleached with the same chemicals used to bleach virgin pulp, but hydrogen peroxide and sodium hydrosulfite are the most common bleaching agents. Recycled pulp, or paper made from it, is known as PCF (process chlorine free) if no chlorine-containing compounds were used in the recycling process.

Recycled Paper and Paper Mills

Recycling as an alternative to the use of landfills and recycled paper is one of the less complicated procedures in the recycling industry. Although there is not a landfill crisis at this point in time, it is commonly believed that measures should to be taken in order to lower the negative impacts of landfills, for many hazardous elements are produced and spread because of this enclosure of trash. Most recycled paper is priced higher than freshly made paper, and this tends to plays a deciding factor for the consumer.

Because most of the recycled pulp is purchased in an open market, virgin paper is produced cheaper with the pulp that was made by the specific paper mill.

Virgin paper contains no recycled content and is made directly from the pulp of trees or cotton. Materials recovered after the initial paper manufacturing process are considered recycled paper.

Because that original standard was so vague, some "recycled papers" contained only mill scraps that would have been included in virgin paper anyway. Standards have recently been set to prevent companies from making it seem like they were selling recycled paper. The collection and recycling industries have fixated on the scraps of paper that is thrown away by customers daily in order to increase the amount of recycled paper. Different paper mills are structured for different types of paper, and most "recovered office paper can be sent to a deinking mill". A deinking mill serves as a step in the recycling paper process.

This type of mill detaches the ink from the paper fibers, along with any other excess materials which are also removed from the remaining paper. In the deinking mill, after all of the unwanted coatings of paper are stripped, the refurbished paper is sent to the paper machine. The old scraps are now constructed into new paper at the paper machine. Many papers mills have recycled business papers by transforming the old business papers into beneficial letters and envelopes.

The production process for recycled paper is more costly than the well-developed paper mills that create paper with the use of trees. This process in making recycled paper is also much more time-consuming. However, recycling paper has a multitude of benefits from an environmental perspective. "For all the state-of-the-art technology now incorporated into modern paper mills, the industry's underlying structure is still based upon a worldview that was transformative in the 19th-century but is out-of-date as the 21st century approaches".

Regulations

"The Environmental, Health, and Safety (EHS) Guidelines are technical reference documents with general and industry specific examples of Good International Industry Practice (GIIP)." In simpler terms, the EHS is what helps develop the federal regulations on industries and companies that require factories that have the potential to cause a great amount of harm to society and the environment. These Guidelines for the environment, health and safety list out the specific rules for the paper mill industries that explains what they need to follow in order to limit the pollution that is consequently distributed and by the mills.

Mechanical pulp mills

Wood pulp produced primarily by grinding wood is known as "mechanical pulp" and is used mainly for newsprint. These mechanical processes use fewer chemicals than either kraft or sulfite mills. The primary source of pollution from these mills is organic material such as resin acids released from the wood when it is processed. Mechanical wood pulp is "brightened," as opposed to bleached, using less toxic chemicals than are needed for chemical pulps.

INKS

Three main issues with the environmental impact of printing inks is the use of volatile organic compounds, heavy metals and non-renewable oils. Standards for the amount of heavy metals in ink have been set by some regulatory bodies. There is a trend toward using vegetable oils rather than petroleum oils in recent years due to a demand for better sustainability.

Deinking recycled paper pulp results in a waste slurry which may go to landfill. De-inking at Cross Pointe's Miami, Ohio mill in the United States results in sludge weighing 22% of the weight of wastepaper recycled.

In the 1970s federal regulations for inks in the United States governed the use of toxic metals such as lead, arsenic, selenium, mercury, cadmium and hexavalent chromium.

INDIAN PAPER INDUSTRY

The Indian Paper Industry accounts for about 2.6% of the world's production of paper. The estimated turnover of the industry is INR 50,000 crore approximately and its contribution to the exchequer is around INR 4500 crore. The industry provides employment to more than 0.5 million people directly and 1.5 million people indirectly.

Most of the paper mills are in existence for a long time and hence present technologies fall in a wide spectrum ranging from oldest to the most modern. The mills use a variety of raw material viz. wood, bamboo, recycled fibre, bagasse, wheat straw, rice husk, etc.; approximately 31% are based on chemical pulp, 47% on recycled fibre and 22% on agro-residues. The geographical spread of the industry as well as market is mainly responsible for regional balance of production and consumption. The operating capacity of the industry currently stands at 12.75 million tons. During 2013-14, domestic production is estimated to be 11.38 million tons. As per industry guesstimates, over all paper consumption has now touched 13.10 million tons and per capita consumption is pegged at 10 kg.

Demand of paper has been hovering around 8% for some time. So far, the growth in paper industry has mirrored the growth in GDP. India is the fastest growing market for paper globally and it presents an exciting scenario; paper consumption is poised for a big leap forward in sync with the economic growth and is estimated to touch 13.95 million tons by 2015-16. The futuristic view is that growth in paper consumption would be in multiples of GDP and hence an increase in consumption by one kg per capita would lead to an increase in demand of 1 million tons.

GROWTH PANGS

The increasing demand for paper brings with it new challenges of economies of scale, efficient usage of resources, need to develop and expand sustainable use of fibre, and value chain management, etc. Despite the fact that the Indian Paper Industry holds its importance to the national economy, unfortunately it stands fragmented.

Paper sector is dominated by small and medium size units; number of mills of capacity 50000 tons per annum or more is not more than 25. Less than half a dozen mills account for almost 90% production of newsprint in the country. There is a growing need to modernize the Indian mills, improve productivity and build new capacities.

AGRO FORESTRY

In India, the paper industry is primarily rural based with close linkages with farming community. Over the years it has evolved into an agro-based industry from its earlier character of a forest-based industry.

Myth and The Reality: While all pervading myth that continues to slur the image of the industry is that it perpetually uses forest raw material and consequently denudes natural forests of the country, the truth is that over the last decade, industry led farm/social forestry have brought around 0.5 million hectares under pulp wood plantations, mainly degraded marginal lands of farmers. At the current estimate, wood based segment of the industry uses 80 per cent of the total requirement from farm produced wood. In particular, IPMA member mills have been actively promoting agro forestry with private land holders/farmers to meet imminent raw material needs in a sustainable manner thereby, positively impacting the Greening India mission of the Government.

Also, the industry initiative has created major employment in the remote areas in close proximity to the manufacturing facilities thereby, helping rural development.

Global Scenario: Globally, paper industry has realized a linkage between production of paper and the farming community thereby infusing huge capital in agricultural economy. The USA, Latin America, Scandinavian countries, Australia, Japan and neighboring Asian countries viz. China, Indonesia etc. all have been quick to create conducive land use/forestry policies to encourage large-scale production plantations and attract mega investments in pulp and paper/paperboard manufacturing and in the process creating millions of jobs.

India Imperative: In order to replicate success stories of major global players in the field of Production/ Industrial Plantation, IPMA continues to advocate for atleast 2% access to degraded forestland on lease. It is IPMA's firm belief that besides facilitation of social/farm forestry on private/revenue waste lands, if paper mills are permitted to grow pulpable wood in the radius of 150- 200 KMs to their facilities on the available degraded forest land as well, it would hugely help mills to achieve cost competitiveness in terms of wood and freight and would also, help generate employment, create wealth for the rural poor and lend protection to environment. In this context, IPMA has pleaded for adoption of Multi Stakeholder Partnership (MSP) model propounded by MoEF for afforestation of degraded land.

ISSUES AND CONCERNS

The Indian paper and paperboards industry has potential and also capabilities to service the growing demand in domestic and international market and also to create huge employment avenues in the rural-India through agro/ production forestry and at mills, provided the competitiveness of the value chain is encouraged by the government.

Major Deterrent – Fibre Deficiency

Wood: India's wood resources are limited therefore, cost of wood is much higher in global comparison. Since there is conspicuous absence of Government's enabling policies favouring industrial/production plantation, securing future wood supplies will be Industry's biggest challenge. Wood based segment of the paper industry meets its current wood requirements mainly through social/farm forestry and supplements with purchases made from the State Forest Development Corporations.

Bagasse/ Straw: Though annual availability of agro residues is large yet, this may not be able to sustain the future growth of the Industry, taking due account of quality of paper required, environmental issues involved, etc. Moreover, bagasse is increasingly used by sugar mills for co-generation of power and no more easily available to the paper mills as raw material.

Waste Paper: Recovered fibre consumption is going up globally. In India about 850,000-1,000,000 tons of waste paper is being currently recovered annually. The recovery rate works out to about 20% which is much lower in comparison with 65% recovery achieved by many global players. Low recovery is on account of alternate use of paper in wrapping, packing, etc. The utilization rate of recovered fibre is only 47%. Paper mills are heavily dependent on imported waste paper which commands exorbitant price due to inadequate availability. India needs a well-defined and aggressive system for collection, sorting, grading and utilization recyclable waste paper to contain imports.

Energy Cost: The Government of India has recently withdrawn core sector status hitherto enjoyed by the paper industry. Cost of coal is escalating and prospect of availability of quality coal is diminishing. The steep price rise in coal has resulted in escalation of cost of production of those mills which happened to be dependent on imported coal for generation of steam/power. Also, power purchased from the grid is proving expensive for the industry.

Certification: Forest stewardship council certification is becoming a non trade barrier for Indian paper companies. As bulk of the raw material is obtained from farm and agro- forestry, IPMA is of the view that it would not be practical for huge number of farmers involved in social/farm forestry to group and obtain the FSC certificate for their produce. Though the farm forestry is a sustainable model promoted by the paper industry, the FSC principles and criterions are difficult to satisfy. IPMA is monitoring the move set afoot by the Government of India to establish Indian Forest Stewardship Council to help the process of certification.

GROWTH AND DISTRIBUTION OF PAPER INDUSTRY

Paper is one of the core industries and is linked to the basic human needs. Paper is the pre-requisite for education and literacy and its use is an index of advancement in these two fields as well as the overall well being of the society.

This is the most important of all the forest based industries. Some people treat it as a chemical industry due to its manufacturing process and because of certain chemicals used for its manufacturing.

Still some other people include it in the group of agro-based industries because some of the agricultural products and residuals are used as raw materials. As large proportion of the basic raw materials is derived from the forests, it seems logical to treat it as a forest based industry.

GROWTH AND DEVELOPMENT:

Paper manufacturing has been carried on in India since tenth century as a small cottage industry by the traditional craftsmen called kagzis. They used gunny bags, rags, ropes, etc. for making paper. This industry could not survive the onslaught of the machine made paper and declined considerably. However, a part of it has managed to survive and even today, a large number of small units are producing handmade paper.

The beginning of modern paper industry goes back to 1816 when a factory was set up near Chennai. This venture proved abortive. Another papermill was set up in 1832 at Serampore on the bank of Hugli in West Bengal. This venture also failed and the first successful effort was made in 1870 with the setting up of the Royal Bengal Paper mills at Ballyganj near Kolkata.

This is the principal seat of paper industry even today. Subsequent successful efforts were made at Lucknow in 1879, Titagarh in 1882, Pune in 1887, Raniganj in 1892, Kaukinra in 1892 and Naihati in 1918. The preferential treatment and tariff protection helped in solving the initial problems of this industry.

World War II further infused life into this industry. But the progress was rather sluggish till Independence and it is only in the plan period that the industry has really made fast progress. In the year 1950-51, there were 17 mills with a total installed capacity of 1.37 lakh tonnes.

The corresponding figures reached 77 mills and 14.40 lakh tonnes in 1977-78 and 379 mills and 37.78 lakh tonnes in 1995-96. Currently, India has a total installed capacity of 6.2 million tonnes of paper and paper board. The aggregate installed capacity by 2010 for paper and paperboard is expected to touch 8.3 million tonnes. The production of paper and paper board has also increased gradually but uninterruptedly from a mere 116 thousand tonnes in 1950-51 to 3,684 thousand tonnes in 2003-04.

RAW MATERIALS:

Paper and paper board can be manufactured by using different types of raw materials. Raw materials account for 45-50 per cent of the total cost of production and form an important segment of manufacturing paper and paper board. Of the total installed capacity, 43 per cent is dependent on forest based raw materials, 28 per cent on agro based raw materials and the remaining 29 per cent on other materials including waste paper.

BAMBOO:

Generally speaking 2.3 to 2.4 tonnes of bamboo is required for producing one tonne of paper. The paper industry uses bamboo to the extent of 60-70 per cent of the total requirements of cellulosic raw materials. Bamboo has the advantage of the possessing long fibre, dense stands and quick regeneration.

It reaches maturity in 2-3 years and provides continuous flow of renewable source of raw material. However, there is danger of this source of important raw material being depleted if the rate of exploitation exceeds the rate of regeneration. The total supply of bamboo at the current rate is estimated at 20-30 lakh tonnes per annum. Assam, Orissa, Andhra Pradesh, Madhya Pradesh,

Tamil Nadu, Karnataka and Maharashtra are important producers of bamboo.

SABAI GRASS:

This is another important raw material for manufacturing paper. It was the sole raw material before introduction of bamboo as a significant raw material, but its use has decreased considerably since then. It now constitutes 7 to 9 per cent of the total cellulosic raw material in the country.

Although sabai grass has long fibre and requires low chemical consumption, it grows in tufts intermixed with other vegetation and it is often difficult to separate impurities from it. Moreover, its supplies are much less than those of bamboo. The annual supply of sabai grass along with other allied grasses is about one million tonnes. It mainly grows in the sub-Himalayan tracts of Shiwaliks and Tarai area.

BAGASSE:

It is a fibrous residue of the sugarcane stalk, mainly from the sugar mills, obtained after sucrose is extracted by crushing the sugarcane. On an average 50-60 lakh tonnes of bagasse is produced in the country, half of which is used for manufacturing paper.

OTHER MATERIALS:

Paper is also manufactured by using materials other than those mentioned above. These include waste paper, rags, straw from rice and wheat, jute sticks and soft wood obtained from eucalyptus, pine-wood, wattle and mulberry trees.

CHEMICALS:

In addition to the above mentioned cellulosic raw materials, certain chemicals are also used in the manufacture of pulp, paper and paper board. The important chemicals are caustic soda, soda ash, sodium sulphate, chlorine, sulphuric acid, sulphur, lime, ferric alumina, ammonium sulphate, resin and clay.

A variety of sizing and colouring agents are also used. Coal is also an important input in paper industry. Normally, 3.5 to 4.1 tonnes of coal is required for producing one tonne of paper. Large quantity of soft water is also needed.

LOCALISATION OF THE INDUSTRY:

Paper and paper board manufacturing uses coarse, cheap and weight losing raw materials and seeks raw material oriented locations. Chemicals used in this industry are needed in small quantity and are easily transported even over long distances from the place of production to the place of consumption.

Therefore, there is a strong tendency among the paper mills to be located near the forest tracts along the Western Ghats, the Eastern Ghats, central India and the Tarai-Bhabar area at the foot hills of the Himalayas. There are vast stands of temperate forests in the Himalayan region which can provide large quantities of soft woods.

These soft woods can provide excellent cellulosic material tor the paper industry. But because of the rugged terrain and poor means of transportation, these areas are inaccessible and the possibility of extraction of these soft woods and their supply to paper-mills in the adjoining areas is remote.

Moreover, the regeneration of wood pulp forests takes about 50-60 years, while bamboo forest is ready within 2-3 years. So far, no paper mill has been set up in the Himalayan region. Since the supply of cellulosic raw materials falls short of the demands of the paper industry, efforts are being made to bring more land under plantation of eucalyptus and other fast growing sort wood trees. Some of the paper mills are located near the market where cheap labour is also available.

DISTRIBUTION OF PAPER INDUSTRY:

The paper industry distribution scenario has undergone some changes during the last few years.

Maharashtra:

Maharashtra has risen from fourth position in the mid-1980s to second position in 1993 and to first position in 1996 among the major paper producing states of India.

The state has 63 mills, accounts for 16.52 per cent of the installed capacity and produces 18 per cent of the paper produced in India. Maharashtra is not very lucky with respect to the availability of raw material and coal.

Most of the mills located in Maharashtra use rags, waste paper or pulp imported from Sweden and Canada as the principal cellulosic raw material. Hydroelectricity is used as a power resource in place of coal. Ballarpur has the largest paper mill. Sangli, Kalyan, Mumbai, Pune, Balarshah, Pimpn, Nagpur, Bhiwandi, Nandurbar, Tumur, Khopoli, Kamptee, Vikroli, Chinchwad, etc. are other major producers.

Andhra Pradesh:

Although a late entrant in paper industry, Andhra Pradesh superceded West Bengal in early 1990s to become the top producer but conceded this position to Maharashtra in mid- 1990s. At present, this state has 19 mills, accounts for 11.3 per cent of installed capacity and 13 per cent of India's total production of paper.

Paper mills are mainly located at Rajahmundry, Sirpur (Kagaznagar), Tirupati, Kurnool, Khammam, Srikakulum, Pallancheru, Nellore Bhadrachalam, Kakinada, Apidik, Bodhan, etc. Most of the mills use bamboo as the basic raw material. Bamboo grows widely in Andhra Pradesh and the recoverable reserves are estimated to be about 1.5 lakh tonnes per annum.

Gujarat:

Gujarat has improved its position considerably in paper manufacture. The share of Gujarat in the nation's total production of paper has gone up from a miserable 3 per cent in the mid- 1980s to a respectable 10 per cent in the mid-1990s.

The state has as many as 55 mills which largely depend upon bagasse and eucalyptus for supply of raw material. The main centres of production are Rajkot, Vadodara, Surat, Barjod, Bilmoria, Navsari, Songarh, Ahmedabad, Vapi, Bharuch, Dijandranagar, Limbdi, Gondal, Udvada and Bavla.

Uttar Pradesh:

This state has the largest number of 68 mills, but the size of the mills being small, the installed capacity does not exceed 9 per cent. Saharanpur and Lalkuan have mills of large size. The other centres are at Meerut, Modinagar, Ghaziabad, Lucknow, Gorakhpur, Pipraich, Muzaffamagar, Allahabad (Naini), Varanasi, Kalpi, Budaun and Mainpuri.

West Bengal:

West Bengal was the pioneer state in the paper industry at the initial stage and led the country till mid-1980s. The state has since slide down the ladder to fifth place with 22 mills, 7 per cent of the installed capacity and about the same percentage of production. West Bengal has the advantage of early start of this industry.

The paper industry in West Bengal is based on bamboo which is available locally or is obtained from Assam, Orissa and Jharkhand, and sabai grass which is obtained from Chhattisgarh and Madhya Pradesh. Coal is abundantly available from Jharia and Raniganj.

Ganga and other rivers provide sufficient water for processing. High density of population provides cheap labour.The highest concentration of paper mills is found along the Hugli River. Titagarh, Kankinara, Raniganj, Bansberia, Sheoraphuli, Chandrabati, Triveni, Naihati, Kolkata and Baranagore are some of the important centres of paper manufacturing in West Bengal.

Madhya Pradesh:

Madhya Pradesh has large tracts under cellulosic raw materials viz., bamboo, sabai grass, eucalyptus, etc. and provides solid base to paper industry. The state has 18 mills which account for 6.62 per cent of the total installed capacity of India. The main centres of production are Bhopal, Amlpi, (Shahdol), Ratlam, Rajgarh, Vidisha, Abdullaganj, Rewa and Indore. Nepanagar is the birthplace of newsprint indun 6 and is still an important centre of newsprint manufacturing in India.

Orissa:

The paper mills of Orissa make use of bamboo which covers large areas in the Eastern Ghats. The average annual production of bamboo is about five lakh tonnes.

The state has only 8 mills but their size is sufficiently large enabling the state to account for over six per cent of the total capacity. However, with the dwindling supply of bamboo, the state's relative position has declined over the years. Brijrajnagar, Chandwar, Raygada are the main centres of production.

Tamil Nadu:

Tamil Nadu has 24 small sized mills accounting for 5.74 per cent of the total installed capacity of the country. These mills use locally grown bamboo. Cheranmahadevi, Pallipalayam, Udmalpet, Chennai, Salem, Amravathinagar, Pahanasam, Madurai, etc. are the major producers.

Karnataka:

Paper mills in Karanataka use locally grown bamboo and bagasse obtained from the sugar mills. Karnataka has 17 mills accounting for 5.48 per cent of the total capacity of India. The main cocentration is found at Bhadravati, Dandoli, Nandangaud, Belagola, Munirabad, Harihar, Mundyud, Bangalore, Mandya, Ramnagaram and Krishnarajsagar.

Punjab:

Punjab has 23 mills but their size is small as compared to those of Karnataka and their installed capacity is a little over five per cent of the total capacity of India. Most of the mills have capacities varying from 10 to 20 thousand tonnes. Hoshiarpur, Sangrur, Sailakhurd and Rajpura are the chief producing centres.

Assam:

Assam has large areas under bamboo growth and supplies bamboo to neighbouring states, like West Bengal after meeting its own requirements. Nowgaon has one of the largest paper mills of India. The other producing centres are Guwahati, Cachar and Lumding.

Haryana:

Haryana's 18 mills mainly depend upon imported pulp and eucalyptus wood as raw material. Yamunanagar has the largest mill. Small units are functioning at Faridabad, Dharuhera and Jagadhari.

Others:

Bihar has 6 mills with a total installed capacity of 87.6 thousand tonnes. Dalmianagar, Patna, Darbhanga, Samastipur, Barauni are the chief producers. Himachal Pradesh also has 6 small units. Borokwala and Kala Amb are the main centres. Kerala (Punalur, Ernakulam, Mavoor, Rayanpuram and Kozhikode), Rajasthan (Kota), Meghalaya (Shillong), Nagaland (Mokekchong) also produce paper.

Newsprint:

The production of newsprint in India commenced with the establishment of National Newsprint and Paper Mills known as NEPA in 1955 at Nepanagar in Hoshangabad district of M.P. Its initial capacity was 30 thousand tonnes which has now been raised to 75 thousand tonnes.

This mill remained as the sole newsprint producer in the country till 1980. Mysore Paper mills, at Shimoga in Karnataka (capacity 75,000 tonnes) started production in 1981. The Hindustan Newsprint at Vellore near Kottayam in Kerala (capcity 80,000 tonnes) started producing in 1982.

This was followed by commencement of production in 1985 by Tamil Nadu Newsprint and Paper Ltd. at Pugalur in Tiruchirapalli with a capacity of 50,000 tonnes. As on 1 January 1996, there were 26 newsprint mills the country with an annual installed capacity of 6.4 lakh tonnes. The current installed capacity is 1.24 million tonnes which is likely to reach a level of 1.5 million tonnes in 2010.

FOREIGN TRADE:

In spite of the phenomenal progress made by the paper industry in India, the production of paper, paper board and above all newsprint has always fallen short of the demand. This has forced the country to resort to heavy imports to meet the growing demand in the domestic market and put a check on the rising price of paper and paper products.

The present rate of consumption of about 5.5 kg per capita per annum is extremely low as compared to the world average of 50 kg and over 300 kg of the highly advanced countries. A little increase in the capita consumption can lead to serious crisis of paper availability in the coming years.

Problems and Prospects:

Paper industry in India faces many serious problems and prospects do not seem very bright. The biggest problem faced by this industry is the scarcity of raw materials. Most of the materials used for manufacturing paper are derived from the forests. With the increasing degradation of forest and fast depletion of forest based raw materials, like bamboo, the paper industry is facing a severe raw material crisis. With the exploitation of forests based raw materials reaching its saturation, any substantial capacity expansion in the near future is practically ruled out and the growth of industry has come to a standstill. In order to survive and expand, the industry will have to look for unconventional raw materials. This will require new advanced technology which a developing country like India may not be able to afford.

Another alternative is to increase the use as waste paper which will reduce dependence on the conventional raw materials. This will also reduce cost of energy used and other inputs. It may be mentioned here that only 15 per cent of the total output of paper and paper board is based on recycled material against the world average of 30-35 per cent. Thus there is a vast scope for using recycled material in paper industry.

Even now, large quantities of bagasse are just used as fuel in the sugar industry and is not made available to paper industry. If the sugar mills are encouraged to use coal fired boilers instead of those based on bagasse, this precious raw material can solve the problem of paper industry to great extent.

The growing consciousness for preservation of forests and maintenance of ecological balance and biodiversity during the last few years is further reducing the availability of raw materials to this industry. Environmentalists are also up in the arms against this industry due to effluents thrown by the paper mills into open drains, rivulets and rivers thereby polluting the environment.

Unless technology to solve the effluent problem is evolved and implemented with only marginal additional investment, many of the paper mills run the risk of economic losses and even closure.

India is facing wide gap between supply and demand of paper even at the low rate of consumption. Currently, India has over 16 per cent of the world's population but consumes a meagre 1 per cent of the world's paper and paper board.

With the spread of education and literacy, demand for paper is bound to increase and is expected to be double the present demand within ten years. These will constraint India either to increase the indigenous production considerably or resort to large scale imports.

The average size of paper mills is abnormally low at less than 10,000 tonnes against 50,000 tonnes in South-east Asia and 85,000 tonnes in Asia Pacific. The small size of the manufacturing units makes them uneconomic, prohibits the induction of new technology, needs larger capital inputs in proportion to the final production and increases the cost of operations.

Several mini plants are using old technologies and worn out machinery and badly lack the competitive strength. The spiralling cost of imported wood pulp and waste paper coupled with unremunerative selling price of the finished paper has rendered many paper mills economically unviable and forced them to close down.

Recently, the Department of Industrial Policy and Promotion had commissioned a study on 'Global Competitiveness' of Indian Paper Industry by an agency of international repute to understand various issues concerning the paper and newsprint industry.

The study had indicated inadequate availability of good quality of cellulosic raw material and obsolete technology, among others as the constraints for the paper industry. High cost of basic inputs and environmental issues are the other two major problems required to be addressed by the paper industry to become globally competitive.

However, one encouraging aspect is that Indian paper industry has trained manpower whose skill can be gainfully used to adopt modern technology for manufacturing paper and paper board of international standard at relatively lower manpower cost inputs.

The country has vast reservoir of unemployed rural labour which can be deployed for development of raw materials. Thus, we can overcome some of the weaknesses which have plagued the paper industry for a long time.

PAPER ENGINEERING

Paper engineering is a branch of engineering that deals with the usage of physical science (e.g. chemistry and physics) and life sciences (e.g. biology and biochemistry) in conjunction with mathematics as applied to the converting of raw materials into useful paper products and co-products. The field applies various principles in process engineering and unit operations to the manufacture of paper, chemicals, energy and related materials. The following timeline shows some of the key steps in the development of the science of chemical and bioprocess engineering:

From a heritage perspective, the field encompasses the design and analysis of a wide variety of thermal, chemical and biochemicalunit operations employed in the manufacture of pulp and paper, and addresses the preparation of its raw materials from trees or other natural resources via a pulping process, chemical and mechanical pretreatment of these recovered biopolymer (e.g. principally, although not solely, cellulose-based) fibers in a fluid suspension, the high-speed forming and initial dewatering of a non-woven web, the development of bulk sheet properties via control of energy and mass transfer operations, as well as post-treatment of the sheet with coating, calendering, and other chemical and mechanical processes.

APPLICATIONS

Today, the field of paper and chemical engineering is applied to the manufacture of a wide variety of products. The (bio)chemical industry scope manufactures organic andagrochemicals (fertilizers, insecticides, herbicides), oleochemicals, fragrances and flavors, food, feed, pharmaceuticals, nutraceuticals, chemicals, polymers and power from biological materials. The resulting products of paper engineering including paper, cardboard, and various paper derivatives are widely used in everyday life. In addition to being a subset of chemical engineering, the field of paper engineering is closely linked to forest management, product recycling, and the mass production of paper – based media.

METHODS

Mechanical pulping

In the process of mechanical pulping, "grinding" and "refining" are the two main methods used to create the pulp. Grinding is the method of pressing logs and chips against a turning stone to produce fibers. Refiner pulping is treating wood chips with chemicals or heat and then crushing the objects between two disks, one or both of which are rotating. There are four main types of refiner pulping, which includes refiner mechanical pulping, thermo-mechanical pulping, chemi-mechanical pulping, and chemithermomechanical pulping. Further descriptions of each process are contained in this link: Mechanical pulping, when compared to chemical pulping, is relatively inexpensive and has a high pulp yield (85–95%). However, the paper created is generally weak since it retains the lignin.

Chemical pulping

The process of chemical pulping is used to chemically disband the lignin found in the cell walls of the material undergoing the process. After the cellulose fibers are separated from the lignin, a pulp is created which can then be treated to create durable paper, boxes, and corrugated cardboard. Chemical pulping can be characterized by two main methods: sulfate (Kraft process) pulping and sulfite pulping, and these two methods have different benefits. Sulfate pulping can be performed on a wide range of tree varieties and results in the creation of a strong type of paper.

Conversely, sulfite pulping results in a higher volume of pulp which is easier to bleach and process. However, sulfate pulping is more widely used since the product is more durable and the chemicals used in the process can be recovered, thus resulting in minimal environmental pollution.

Further pulp processing

The pulp is then processed through an apparatus which renders the pulp as a mesh of fibers. This fiber network is then pressed to remove all water contents, and the paper is subsequently dried to remove all traces of moisture.

Finishing

After the above processes have been completed, the resulting paper is coated with a minuscule amount of china clay or calcium carbonate to modify the surface, and the paper is then re-sized depending on its intended purpose.

Product recycling

Generally, the material to be recycled first undergoes mechanical or chemical pulping to render it in pulp form. The resulting pulp is then processed in the same way normal pulp is processed; however, original fiber is sometimes added to enhance the quality and appearance of the product.

PAPER RECYCLING

Fig. Waste paper collected for recycling in Italy

There are three categories of paper that can be used as feedstocks for making recycled paper: mill broke, pre-consumer waste, and post-consumer waste. *Mill broke* is paper trimmings and other paper scrap from the manufacture of paper, and is recycled internally in a paper mill. *Pre-consumer waste* is material which left the paper mill but was discarded before it was ready for consumer use. *Post-consumer* waste is material discarded after consumer use, such as old corrugated containers (OCC), old magazines, and newspapers.Paper suitable for recycling is called "scrap paper", often used to produce molded pulp packaging. The industrial process of removing printing ink from paperfibers of recycled paper to make deinked pulp is calleddeinking, an invention of the German jurist Justus Claproth.

PAPER RECYCLING PROCESS

The process of waste paper recycling involves mixing used paper with water and chemicals to break it down. It is then chopped up and heated, which breaks it down further into strands of cellulose, a type of organic plant material; this resulting mixture is called pulp, or slurry. It is strained through screens, which remove any glue or plastic that may still be in the mixture then cleaned, de-inked, bleached, and mixed with water. Then it can be made into new recycled paper.

The share of ink in a wastepaper stock is up to about 2% of the total weight.

RATIONALE FOR RECYCLING

Industrialized paper making has an effect on the environment both upstream (where raw materials are acquired and processed) and downstream (waste-disposal impacts).

Today, 40% of paper pulp is created from wood (in most modern mills only 9-16% of pulp is made from pulp logs; the rest comes from waste wood that was traditionally burnt). Paper production accounts for about 35% of felled trees, and represents 1.2% of the world's total economic output. Recycling one ton of newsprint saves about 1 ton of wood while recycling 1 ton of printing or copier paper saves slightly more than 2 tons of wood.

This is because kraft pulping requires twice as much wood since it removes lignin to produce higher quality fibres than mechanical pulping processes. Relating tons of paper recycled to the number of trees not cut is meaningless, since tree size varies tremendously and is the major factor in how much paper can be made from how many trees.

Trees raised specifically for pulp production account for 16% of world pulp production, old growth forests 9% and second- and third- and more generation forests account for the balance. Most pulp mill operators practice reforestation to ensure a continuing supply of trees.

The Programme for the Endorsement of Forest Certification (PEFC) and the Forest Stewardship Council (FSC) certify paper made from trees harvested according to guidelines meant to ensure good forestry practices. It has been estimated that recycling half the world's paper would avoid the harvesting of 20 million acres (81,000 km^2) of forestland.

Energy

Energy consumption is reduced by recycling, although there is debate concerning the actual energy savings realized. The Energy Information Administration claims a 40% reduction in energy when paper is recycled versus paper made with unrecycled pulp, while the Bureau of International Recycling (BIR) claims a 64% reduction. Some calculations show that recycling one ton of newspaper saves about 4,000 kWh (14 GJ) of electricity, although this may be too high. This is enough electricity to power a 3-bedroom European house for an entire year, or enough energy to heat and air-condition the average North American home for almost six months. Recycling paper to make pulp actually consumes more fossil fuels than making new pulp via the kraft process; these mills generate most of their energy from burning waste wood (bark, roots, sawmill waste) and byproduct lignin (black liquor). Pulp mills producing new mechanical pulp use large amounts of energy; a very rough estimate of the electrical energy needed is 10 gigajoules per tonne of pulp (2500 kW·h per short ton).

Landfill use

About 35% of municipal solid waste (before recycling) by weight is paper and paper products.

Water and air pollution

The United States Environmental Protection Agency (EPA) has found that recycling causes 35% less water pollution and 74% less air pollution than making virgin paper. Pulp mills can be sources of both air and water pollution, especially if they are producing bleached pulp. Modern mills produce considerably less pollution than those of a few decades ago. Recycling paper decreases the demand for virgin pulp, thus reducing the overall amount of air and water pollution associated with paper manufacture. Recycled pulp can be bleached with the same chemicals used to bleach virgin pulp, but hydrogen peroxide and sodium hydrosulfite are the most common bleaching agents. Recycled pulp, or paper made from it, is known as PCF (process chlorine free) if no chlorine-containing compounds were used in the recycling process. However, recycling mills may have polluting by-products like sludge. De-inking at Cross Pointe's Miami, Ohio mill results in sludge weighing 22% of the weight of wastepaper recycled.

RECYCLING FACTS AND FIGURES

In the mid-19th century, there was an increased demand for books and writing material. Up to that time, paper manufacturers had used discarded linen rags for paper, but supply could not keep up with the increased demand. Books were bought at auctions for the purpose of recycling fiber content into new paper, at least in the United Kingdom, by the beginning of the 19th century.

Internationally, about half of all recovered paper comes from converting losses (pre-consumer recycling), such as shavings and unsold periodicals; approximately one third comes from household or post-consumer waste.

Some statistics on paper consumption:

- In 1996 it was estimated that 95% of business information is still stored on paper.
- Recycling 1 short ton (0.91 t) of paper saves 17 mature trees, 7 thousand US gallons (26 m) of water, 3 cubic yards (2.3 m^3) of landfill space, 2 barrels of oil (84 US gal or 320 l), and 4,100 kilowatt-hours (15 GJ) of electricity – enough energy to power the average American home for six months.
- Although paper is traditionally identified with reading and writing,

communications has now been replaced by packaging as the single largest category of paper use at 41% of all paper used.

- 115 billion sheets of paper are used annually for personal computers. The average web user prints 16 pages daily.
- Most corrugated fiberboard boxes have over 25% recycled fibers. Some are 100% recycled fiber.
- Recycling one ton of paper saves roughly 17 trees.
- In 1997, 299,044 metric tons of paper was produced (including paperboard).
- In the United States, the average consumption of paper per person in 1999 was approximately 354 kilograms. This would be the same consumption for 6 people in Asia or 30 people in Africa.
- In 2006-2007, Australia 5.5 million tonnes of paper and cardboard was used with 2.5 million tonnes of this recycled.
- Newspaper manufactured in Australia has 40% recycled content.

PAPER RECYCLING BY REGION

European Union

Paper recovery in Europe has a long history and has grown into a mature organization. In 2004 the paper recycling rate in Europe was 54.6% or 45.5 million short tons (41.3 Mt). The recycling rate in Europe reached 64.5% in 2007, leaving the industry on track to meeting its voluntary target of 66% by 2010.

Japan

Municipal collections of paper for recycling are in place. However, according to the *Yomiuri Shimbun*, in 2008, eight paper manufacturers in Japan have admitted to intentionally mislabeling recycled paper products, exaggerating the amount of recycled paper used.

United States

Recycling has long been practiced in the United States. In 2012, paper and paperboard accounted for 68 million tons of municipal solid waste generated in the U.S., down from more than 87 million tons in 2000, according to the U.S. Environmental Protection Agency. While paper is the most commonly recycled material - 64.6 percent was recovered in 2012 — it is being used less overall than at the turn of the century. Paper accounts for more than a half of all recyclables collected in the US, by weight.

The history of paper recycling has several dates of importance:

- 1690: The first paper mill to use recycled linen was established by the Rittenhouse family.

- 1896: The first major recycling center was started by the Benedetto family in New York City, where they collected rags, newspaper, and trash with a pushcart.
- 1993: The first year when more paper was recycled than was buried in landfills.

Today, over half of all paper used in the United States is collected and recycled. Paper products are still the largest component of municipal solid waste, making up more than 40% of the composition of landfills. In 2006, a record 53.4% of the paper used in the US (53.5 million tons) was recovered for recycling, up from a 1990 recovery rate of 33.5%. The US paper industry set a goal of recovering 55 percent of all paper used in the US by 2012. Paper products used by the packaging industry were responsible for about 77% of packaging materials recycled, with more than 24 million pounds recovered in 2005.

By 1998, some 9,000 curbside recycling programs and 12,000 recyclable drop-off centers existed nationwide. As of 1999, 480 materials recovery facilities had been established to process the collected materials. Recently, junk mail has become a larger part of the overall recycling stream, compared to newspapers or personal letters. However, the increase in junk mail is still smaller compared to the declining use of paper from those sources.

In 2008, the global financial crisis caused the price of old newspapers to drop in the U.S. from $130 to $40 per short ton ($140/t to $45/t) in October.

Mexico

In Mexico, recycled paper, rather than wood pulp, is the principal feedstock in papermills accounting for about 75% of raw materials.

LIMITATIONS AND IMPACTS OF PAPER RECYCLING

Along with fibres, paper might contain a variety of inorganic and organic constituents. Paper might contain up to 10,000 different chemicals, which can potentially contaminate the newly manufactured paper products. As an example, bisphenol A (a chemical commonly found in thermal paper) has been verified as contaminant in a variety of paper products resulting from paper recycling.

3

Pulp (Paper)

Pulp is a lignocellulosic fibrous material prepared by chemically or mechanically separating cellulose fibres from wood, fiber crops or waste paper. The wood fiber sources required for pulping are "45% sawmill residue, 21% logs and chips, and 34% recycled paper" (Canada, 2014). Pulp is one of the most abundant raw materials worldwide.

Fig. Structure of fibres of pulp

HISTORY

Using wood to make paper is a fairly recent innovation. Paper making using cotton and linen fibres spread to Europe in the 13th century. Medieval historian Lynn White credited the spinning wheel with increasing the supply of rags, which led to cheap paper, which was a factor in the development of printing. By the 1800s, fibre crops such as flax, which provided linen fibres, were still the primary material source, and paper was a relatively expensive commodity. The use of wood to make pulp for paper began with the development of mechanical pulping in Germany by F.G. Keller in the 1840s, and by the Canadian inventor Charles Fenerty in Nova Scotia. Chemical processes quickly followed, first with J. Roth's use of sulfurous acid to treat wood, followed by Benjamin Tilghman's U.S. patent on the use of calcium bisulfite, $Ca(HSO_3)_2$, to pulp wood in 1867.Almost a decade later the first commercial sulfite pulp mill was built in Sweden. It used magnesium as the counter ion and was based on work by Carl Daniel Ekman. By 1900, sulfite pulping had become the dominant means of producing wood pulp, surpassing mechanical pulping methods. The competing chemical pulping process, the sulfate or kraft process was developed by Carl F. Dahl in 1879 and the first kraft mill started (in Sweden) in 1890. The invention of the recovery boiler by G.H. Tomlinson in the early 1930sallowed kraft mills to recycle almost all of their pulping chemicals. This, along with the ability of the kraft process to accept a wider variety of types of wood and produce stronger fibres made the kraft process the dominant pulping process starting in the 1940s.

Global production of wood pulp in 2006 was 175 million tons (160 million tonnes). In the previous year, 63 million tons (57 million tonnes) of market pulp (not made into paper in the same facility) was sold, with Canada being the largest source at 21 percent of the total, followed by the United States at 16 percent. Chemical pulp made up 93 percent of market pulp.

Manufacture of wood pulp

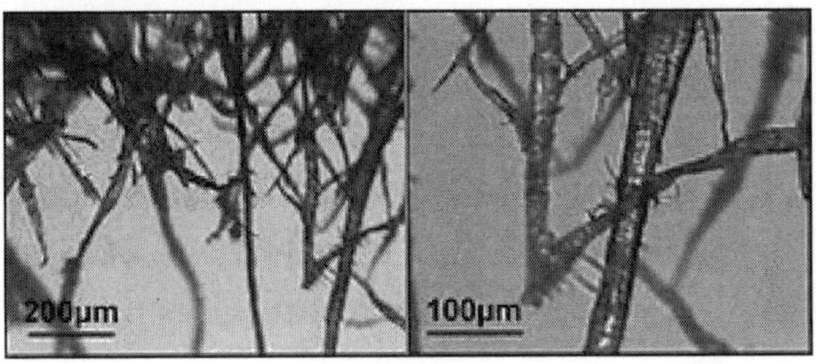

Fig. Fibres in wood pulp

The timber resources used to make wood pulp are referred to as pulpwood. Wood pulp comes from softwood trees such asspruce, pine, fir, larch and hemlock, and hardwoods such as eucalyptus, aspen and birch.

A pulp mill is a manufacturing facility that converts wood chips or other plant fibre source into a thick fiberboard which can be shipped to a paper mill for further processing. Pulp can be manufactured using mechanical, semi-chemical or fully chemical methods (kraft and sulfite processes). The finished product may be either bleached or non-bleached, depending on the customer requirements.

Wood and other plant materials used to make pulp contain three main components (apart from water): cellulose fibers (desired for papermaking), lignin (a three-dimensional polymer that binds the cellulose fibres together) and hemicelluloses, (shorter branched carbohydrate polymers). The aim of pulping is to break down the bulk structure of the fibre source, be it chips, stems or other plant parts, into the constituent fibres.

Chemical pulping achieves this by degrading the lignin and hemicellulose into small, water-soluble molecules which can be washed away from the cellulose fibres without depolymerizing the cellulose fibres (chemically depolymerizing the cellulose weakens the fibres). The various mechanical pulping methods, such as groundwood (GW) and refiner mechanical (RMP) pulping, physically tear the cellulose fibres one from another. Much of the lignin remains adhering to the fibres. Strength is impaired because the fibres may be cut.

There are a number of related hybrid pulping methods that use a combination of chemical and thermal treatment to begin an abbreviated chemical pulping process, followed immediately by a mechanical treatment to separate the fibres. These hybrid methods include thermomechanical pulping, also known as TMP, and chemithermomechanical pulping, also known as CTMP. The chemical and thermal treatments reduce the amount of energy subsequently required by the mechanical treatment, and also reduce the amount of strength loss suffered by the fibres.

Global pulp production by category (2000)

Pulp category	Production [M ton]
Chemical	131.2
-Kraft	117.0
-Sulfite	7.0
-Semichemical	7.2
Mechanical	37.8
Nonwood	18.0
Total virgin fibres	**187.0**
Recovered fibres	147.0
Total pulp	**334.0**

Harvesting trees

Many kinds of paper are made from wood with nothing else mixed into them (with some exceptions, like resume paper, which may include cotton). This includes newspaper, magazines and even toilet paper. Most pulp mills use good forest management practices in harvesting trees to ensure that they have a sustainable source of raw materials. One of the major complaints about harvesting wood for pulp mills is that it reduces the biodiversity of the harvested forest.

Trees raised specifically for pulp production account for 16 percent of world pulp production, old-growth forests account for 9 percent, and second- and third- and more generation forests account for the rest. Reforestation is practiced in most areas, so trees are a renewable resource. The FSC (Forest Stewardship Council), SFI (Sustainable Forestry Initiative), PEFC (Programme for the Endorsement of Forest Certification), and other bodies certify paper made from trees harvested according to guidelines meant to ensure good forestry practices.

The number of trees consumed depends on whether mechanical processes or chemical processes are used. It has been estimated that based on a mixture of softwoods and hardwoods 12 metres (40 ft) tall and 15-20 centimetres (6–8 in) in diameter, it would take an average of 24 trees to produce 0.9 tonne (1 ton) of printing and writing paper, using the kraft process (chemical pulping). Mechanical pulping is about twice as efficient in using trees since almost all of the wood is used to make fibre therefore it takes about 12 trees to make 0.9 tonne (1 ton) of mechanical pulp or newsprint.

There are roughly 2 short tons in a cord of wood.

Preparation for pulping

Wood chipping is the act and industry of chipping wood for pulp, but also for other processed wood products and mulch. Only the heartwood and sapwood are useful for making pulp. Bark contains relatively few useful fibers and is removed and used as fuel to provide steam for use in the pulp mill. Most pulping processes require that the wood be chipped and screened to provide uniform sized chips.

Pulping

There are a number of different processes which can be used to separate the wood fiber:

Mechanical pulp

Manufactured grindstones with embedded silicon carbide or aluminum oxide can be used to grind small wood logs called "bolts" to make stone pulp (SGW). If the wood is steamed prior to grinding it is known as pressure ground wood pulp (PGW). Most modern mills use chips rather than logs and ridged metal discs called refiner plates instead of grindstones. If the chips are just ground up with the plates, the pulp is called refiner mechanical pulp (RMP) and if the chips are steamed while being refined the pulp is called thermomechanical pulp (TMP). Steam treatment significantly reduces the total energy needed to make the pulp and decreases the damage (cutting) to fibres. Mechanical pulps are used for products that require less strength, such as newsprint and paperboards.

Thermomechanical pulp

Thermomechanical pulp is pulp produced by processing wood chips using heat (thus "thermo-") and a mechanical refining movement (thus "-mechanical"). It is a two-stage process where the logs are first stripped of their bark and converted into small chips. These chips have a moisture content of around 25–30 percent and a mechanical force is applied to the wood chips in a crushing or grinding action which generates heat and water vapour and softens the lignin thus separating the individual fibres. The pulp is then screened and cleaned, any clumps of fibre are reprocessed. This process gives a high yield of fibre from the timber (around 95 percent) and as the lignin has not been removed, the fibres are hard and rigid.

Chemi-Thermomechanical pulp

Wood chips can be pretreated with sodium carbonate, sodium hydroxide, sodium sulfite and other chemicals prior to refining with equipment similar to a mechanical mill.

The conditions of the chemical treatment are much less vigorous (lower temperature, shorter time, less extreme pH) than in a chemical pulping process since the goal is to make the fibres easier to refine, not to remove lignin as in a fully chemical process. Pulps made using these hybrid processes are known as chemithermomechanical pulps (CTMP).

Chemical pulp

Chemical pulp is produced by combining wood chips and chemicals in large vessels known as digesters where heat and the chemicals break down the lignin, which binds the cellulose fibres together, without seriously degrading the cellulose fibres. Chemical pulp is used for materials that need to be stronger or combined with mechanical pulps to give a product different characteristics. The kraft process is the dominant chemical pulping method, with the sulfite process being second. Historically soda pulping was the first successful chemical pulping method.

Recycled pulp

Recycled pulp is also called deinked pulp (DIP). DIP is recycled paper which has been processed by chemicals, thus removing printing inks and other unwanted elements and freed the paper fibres. The process is called deinking.

DIP is used as raw material in papermaking. Many newsprint, toilet paper and facial tissue grades commonly contain 100 percent deinked pulp and in many other grades, such as lightweight coated for offset and printing and writing papers for office and home use, DIP makes up a substantial proportion of the furnish.

Organosolv pulping

Organosolv pulping uses organic solvents at temperatures above 140 °C to break down lignin and hemicellulose into soluble fragments. The pulping liquor is easily recovered by distillation. The reason for using a solvent is to make the lignin more soluble in the cooking liquor. Most common used solvents are methanol, ethanol, formic acid and acetic acidoften in combination with water.

Alternative pulping methods

Research is under way to develop biopulping (biological pulping), similar to chemical pulping but using certain species of fungi that are able to break down the unwanted lignin, but not the cellulose fibres. In the biopulping process, the fungal enzyme lignin peroxidase selectively digests lignin to leave remaining cellulose fibres.

This could have majorenvironmental benefits in reducing the pollution associated with chemical pulping. The pulp is bleached using chlorine dioxide stage followed by neutralization and calcium hypochlorite.

The oxidizing agent in either case oxidizes and destroys the dyes formed from the tannins of the wood and accentuated (reinforced) by sulfides present in it.

Steam exploded fibre is a pulping and extraction technique that has been applied to wood and other fibrous organic material.

Bleaching

The pulp produced up to this point in the process can be bleached to produce a white paper product. The chemicals used to bleach pulp have been a source of environmental concern, and recently the pulp industry has been using alternatives to chlorine, such as chlorine dioxide, oxygen, ozone and hydrogen peroxide.

ALTERNATIVES TO WOOD PULP

Today, some people and groups advocate using field crop fibre or agricultural residues instead of wood fibre as being more sustainable.

There is enough straw to meet much of North America's book, magazine, catalogue and copy paper needs. Agricultural-based paper is a guaranteed way to reduce the stress of paper production on old growth and endangered forests. Some agricultural residue pulps take less time to cook than wood pulps. That means agricultural-based paper uses less energy, less water and fewer chemicals. Pulp made from wheat and flax straw has half the ecological footprint of pulp made from forests.

However, wood is also a renewable resource, with about 90 percent of pulp coming from plantations or reforested areas. Non-wood fibre sources account for about 5–10 percent of global pulp production, for a variety of reasons, including seasonal availability, problems with chemical recovery, brightness of the pulp etc. Non-wood pulp processing requires a high use of water and energy.

Nonwovens are in some applications alternatives to paper made from wood pulp, like filter paper or tea bags.

Comparison of typical feedstocks used in pulping		
Component	**Wood**	**Nonwood**
Carbohydrates	65–80%	50–80%
Cellulose	40–45%	30–45%
Hemicellulose	23–35%	20–35%
Lignin	20–30%	10–25%
Extractives	2–5%	5–15%
Proteins	<0.5%	5–10%
Inorganics	0.1–1%	0.5–10%
SiO_2	<0.1%	0.5–7%

MARKET PULP

Market pulp is any variety of pulp that is produced in one location, dried and shipped to another location for further processing. Important quality parameters for pulp not directly related to the fibres are brightness, dirt levels, viscosity and ash content. In 2004 about it accounted for about 55 million metric tons of market pulp.

Air dry pulp

Air dry pulp is the most common form to sell pulp. This is pulp dried to about 10 percent moisture content. It is normally delivered as sheeted bales of 250 kg. The reason to leave 10 percent moisture in the pulp is that this minimizes the fibre to fibre bonding and makes it easier to disperse the pulp in water for further processing to paper.

Roll pulp

Roll pulp or reel pulp is the most common delivery form of pulp to non traditional pulp markets. Fluff pulp is normally shipped on rolls (reels). This pulp is dried to 5–6 percent moisture content. At the customer this is going to a comminution process to prepare for further processing.

Flash dried pulp

Some pulps are flash dried. This is done by pressing the pulp to about 50 percent moisture content and then let it fall through silos that are 15–17 m high. Gas fired hot air is the normal heat source. The temperature is well above the char point of cellulose, but large amount of moisture in the fibre wall and lumen prevents the fibres from being incinerated. It is often not dried down to 10 percent moisture (air dry). The bales are not as densely packed as air dry pulp.

ENVIRONMENTAL CONCERNS

The major environmental impacts of producing wood pulp come from its impact on forest sources and from its waste products.

Forest resources

The impact of logging to provide the raw material for wood pulp is an area of intense debate. Modern logging practices, using forest management seek to provide a reliable, renewable source of raw materials for pulp mills. The practice of clear cutting is a particularly sensitive issue since it is a very visible effect of logging. Reforestation, the planting of tree seedlings on logged areas, has also been criticized for decreasing biodiversity because reforested areas are monocultures. Logging of old growth forests accounts for less than 10 percent of wood pulp, but is one of the most controversial issues.

Effluents from pulp mills

The process effluents are treated in a biological effluent treatment plant, which guarantees that the effluents are not toxic in the recipient.

Mechanical pulp is not a major cause for environmental concern since most of the organic material is retained in the pulp, and the chemicals used (hydrogen peroxide andsodium dithionite) produce benign byproducts (water and sodium sulfate (finally), respectively).

Chemical pulp mills, especially kraft mills, are energy self-sufficient and very nearly closed cycle with respect to inorganic chemicals.

Bleaching with chlorine produces large amounts of organochlorine compounds, including dioxins.

Odor problems

The kraft pulping reaction in particular releases foul-smelling compounds. The hydrogen sulfide reagent that degrades lignin structure also causes some demethylation to produce methanethiol, dimethyl sulfide and dimethyl disulfide. These same compounds are released during many forms of microbial decay, including the internal microbial action in Camembert cheese, although the kraft process is a chemical one and does not involve any microbial degradation. These compounds have extremely low odor thresholds and disagreeable smells; a common saying in communities economically dependent on nearby papermills is "Smells like a job" or "Smells like money."

APPLICATIONS

The main applications for pulp are paper and board production. The furnish of pulps used depends on the quality on the finished paper. Important quality parameters are wood furnish, brightness, viscosity, extractives, dirt count and strength. Chemical pulps are used for making nanocellulose.

Speciality pulp grades have many other applications. Dissolving pulp is used in making regenerated cellulose that is used textile and cellophane production. It is also used to make cellulose derivatives. Fluff pulp is used in diapers, feminine hygiene products and nonwovens.

Paper production

The Fourdrinier Machine is the basis for most modern papermaking, and it has been used in some variation since its conception. It accomplishes all the steps needed to transform a source of wood pulp into a final paper product.

ECONOMICS

In 2009, NBSK pulp sold for $650/ton in the United States. The price had dropped due to falling demand when newspapers reduced their size, in part, as a result of the recession.

LOGGING

Fig. McGiffert Log Loader in East Texas, USA circa 1907

Logging is the cutting, skidding, on-site processing, and loading of trees or logs onto trucks or skeleton cars.

In forestry, the term logging is sometimes used in a narrow sense concerning the logistics of moving wood from the stump to somewhere outside the forest, usually a sawmill or a lumber yard. However, in common usage, the term may be used to indicate a range of forestry or silviculture activities.

Illegal logging refers to what in forestry might be called timber theft by the timber mafia. It can also refer to the harvest, transportation, purchase or sale of timber in violation of laws.

The harvesting procedure itself may be illegal, including using corrupt means to gain access to forests; extraction without permission or from a protected area; the cutting of protected species; or the extraction of timber in excess of agreed limits.

Clearcut logging is not necessarily considered a type of logging but a harvest or silviculture method and is simply called clearcutting or block cutting. In the forest products industry logging companies may be referred to as logging contractors, with the smaller, non-union crews referred to as "gyppo loggers."

Cutting trees with the highest value and leaving those with lower value, often diseased or malformed trees, is referred to as high grading. It is sometimes called selective logging, and confused with selection cutting, the practice of managing stands by harvesting a proportion of trees.

Logging usually refers to above-ground forestry logging. Submerged forests exist on land that has been flooded by damming to createreservoirs. Such trees are logged using underwater logging or by the lowering of the reservoirs in question.

Ootsa Lake and Williston Lake in British Columbia, Canada, are notable examples where timber recovery has been needed to remove inundated forests.

CLEARCUTTING

Clearcutting, or clearfelling, is a method of harvesting that removes essentially all the standing trees in a selected area. Depending on management objectives, a clearcut may or may not have reserve trees left to attain goals other than regeneration, including wildlife habitat management, mitigation of potential erosion or water quality concerns.Silviculture objectives for clearcutting, (for example, healthy regeneration of new trees on the site) and a focus on forestry distinguish it from deforestation. Other methods include shelterwood cutting, group selective, single selective, seed-tree cutting, patch cut and retention cutting.

Logging methods

Fig. The Washington Iron Works Skidderin Nuniong is the only one of its kind in Australia, with donkey engine, spars and cables still rigged for work.

The above operations can be carried out by different methods, of which the following three are considered industrial methods:

Tree-length logging / Stem Only Harvesting (SOH)

Trees are felled and then delimbed and topped at the stump. The log is then transported to the landing, where it is bucked and loaded on a truck. This leaves the slash (and the nutrients it contains) in the cut area where it must be further treated if wild land fires are of concern.

Whole Tree Logging (WTL)

Fig. Cable logging in French Alps (cable grue Larix 3T)

Fig. Logging in Hungary

Trees and plants are felled and transported to the roadside with top and limbs intact. There have been advancements to the process which now allows a logger or harvester to cut the tree down, top, and delimb a tree in the same process. This ability is due to the advancement in the style felling head that can be used. The trees are then delimbed, topped, and bucked at the landing. This method requires that slash be treated at the landing. In areas with access to cogeneration facilities, the slash can be chipped and used for the production of electricity or heat. Full-tree harvesting also refers to utilization of the entire tree including branches and tops. This technique removes both nutrients and soil cover from the site and so can be harmful to the long term health of the area if no further action is taken, however, depending on the species, many of the limbs are often broken off in handling so the end result may not be as different from tree-length logging as it might seem.

Cut-to-length logging

Cut-to-length logging is the process of felling, delimbing, bucking and sorting (pulpwood, sawlog, etc.) at the stump area, leaving limbs and tops in the forest. Harvesters fell the tree, delimb and buck it, and place the resulting logs in bunks to be brought to the landing by askidder or forwarder. This method is routinely available for trees up to 900 mm (35 in) in diameter. Harvesters are employed effectively in level to moderately steep terrain. Harvesters are highly computerized to optimize cutting length, control harvest area by GPS and utilize price list for each specific logs to archive most economical results during harvesting.

TRANSPORTING LOGS

Fig. Log transport in the Dutch East Indes (now Indonesia) c. 1870.

Felled logs are then generally transported to a sawmill to be cut into lumber, to a paper mill for paper pulp, or for other uses, for example, as fence posts. Many methods have been used to move logs from where they were cut to a rail line or directly to a sawmill or paper mill. The cheapest and historically most common method is making use of a river's current to float floating tree trunks downstream, by eitherlog driving or timber rafting. (Some logs sink because of high resin content; these are called deadheads.) To help herd the logs to the mill, in 1960 the Alaskan Lumber and Pulp Mill had a specially designed boat that was constructed of 1 $^{1}D_{2}$ inch (38 mm) steel. In the late 1800s and the first half of the 1900s, the most common method was the high-wheel loader, which was a set of wheels over ten feet tall that the log or logs were strapped beneath. Oxen were at first used with the high-wheel loaders, but in the 1930s tractors replaced the oxen. In 1960 the largest high wheel loader was built for service in California. Called the Bunyan Buggie, the unit was self-propelled and had wheels 24 feet (7.3 m) high and a front dozer blade that was 30 feet (9.1 m) across and 6 feet (1.8 m) high.

Fig. Timberjack Harvester at work

Log transportation can be challenging and costly since trees are often far from roads or watercourses. Road building and maintenance may be restricted in National Forests or other wilderness areas since it can cause erosion in riparian zones.

When felled logs sit adjacent to a road, heavy machinery may simply lift logs onto trucks. Most often, special heavy equipment is used to gather the logs from the site and move them close to the road to be lifted on trucks. Many methods exist to transport felled logs lying away from roads. Cable logging involves a yarder, which pulls one or several logs along the ground to a platform where a truck is waiting.

When the terrain is too uneven to pull logs on the ground, a skyline can lift logs off the ground vertically, similar to a ski lift. Heli-logging, which uses heavy-lift helicopters to remove cut trees from forests by lifting them on cables attached to a helicopter, may be used when cable logging is not allowed for environmental reasons or when roads are lacking. It reduces the level of infrastructure required to log in a specific location, reducing the environmental impact of logging. Less mainstream or now for the most part superseded forms of log transport include horses, oxen, or balloon logging.

SAFETY CONSIDERATIONS

Logging is a dangerous occupation. In the United States, it has consistently been one of the most hazardous industries. In 2008, the logging industry employed 86,000 workers, and accounted for 93 deaths. This resulted in a fatality rate of 108.1 deaths per 100,000 workers that year. This rate is over 30 times higher than the overall fatality rate. Loggers work with heavy, moving weights and the use of tools such as chainsaws and heavy equipment on uneven and sometimes steep or unstable terrain. Loggers also deal with severe environmental conditions, such as inclement weather and severe heat or cold. An injured logger is often far from professional emergency treatment.

Traditionally, the cry of "Timber!" developed as a warning alerting fellow workers in an area that a tree is being felled, so they should be alert to avoid being struck. The term "widowmaker" for timber that is neither standing nor fallen to the ground demonstrates another emphasis on situational awareness as a safety principle.

In British Columbia, Canada, the BC Forest Safety Council was created in September 2004 as a not-for-profit society dedicated to promoting safety in the forest sector. It works with employers, workers, contractors and government agencies to implement fundamental changes necessary to make it safer to earn a living in forestry.

The risks experienced in logging operations can be somewhat reduced, where conditions permit, by the use of mechanical tree harvesters, skidders and forwarders.

KRAFT PROCESS

The kraft process (also known as kraft pulping or sulfate process) is a process for conversion of wood into wood pulpconsisting of almost pure cellulose fibers. It entails treatment of wood chips with a mixture of sodium hydroxide and sodium sulfide, known as white liquor, that breaks the bonds that link lignin to the cellulose.

HISTORY

The kraft process (so called because of the superior strength of the resulting paper, from the German word *Kraft* for 'strength') was invented by Carl F. Dahl in 1879 in Danzig,Prussia, Germany. U.S. Patent 296,935 was issued in 1884, and a pulp mill using this technology started (in Sweden) in 1890. The invention of the recovery boiler by G.H. Tomlinson in the early 1930s was a milestone in the advancement of the kraft process. It enabled the recovery and reuse of the inorganic pulping chemicals such that a kraft mill is a nearly closed-cycle process with respect to inorganic chemicals, apart from those used in the bleaching process. For this reason, in the 1940s, the kraft process superseded the sulfite process as the dominant method for producing wood pulp.

THE PROCESS

Impregnation

Common wood chips used in pulp production are 12–25 millimetres (0.47–0.98 in) long and 2–10 millimetres (0.079–0.394 in) thick. The chips normally first enter the *presteaming* where they are wetted and preheated with steam. Cavities inside fresh wood chips are partly filled with liquid and partly with air. The steam treatment causes the air to expand and about 25% of the air to be expelled from the chips. The next step is to saturate the chips with black and white liquor. Air remaining in chips at the beginning of liquor impregnation is trapped within the chips.

The impregnation can be done before or after the chips enters the digester and is normally done below 100 °C (212 °F). The cooking liquors consist of a mixture of white liquor, water in chips, condensed steam and weak black liquor. In the impregnation, cooking liquor penetrates into the capillary structure of the chips and low temperature chemical reactions with the wood begin. A good impregnation is important to get a homogeneous cook and low rejects. About 40–60% of all alkali consumption in the continuous process occurs in the impregnation zone.

Cooking

The wood chips are then cooked in pressurized vessels called digesters. Some digesters operate in a batch manner and some in a continuous process. There are several variations of the cooking processes both for the batch and the continuous digesters. Digesters producing 1,000 tonnes or more of pulp per day are common, with the largest producing more than 3,500 tonnes per day. In a continuous digester, the materials are fed at a rate which allows the pulping reaction to be complete by the time the materials exit the reactor. Typically, delignification requires several hours at 170 to 176 °C (338 to 349 °F). Under these conditions lignin and hemicellulose degrade to give fragments that are soluble in the strongly basic liquid. The solid pulp (about 50% by weight of the dry wood chips) is collected and washed. At this point the pulp is known as *brown stock*because of its color. The combined liquids, known as black liquor (because of its color), contain lignin fragments, carbohydrates from the breakdown of hemicellulose, sodium carbonate, sodium sulfate and other inorganic salts.

One of the main chemical reactions that underpin the kraft process is the scission of ether bonds by the nucleophilic sulfide ($S^{2''}$) orbisulfide ($HS^{''}$) ions.

Recovery process

The excess black liquor contains about 15% solids and is concentrated in a multiple effect evaporator. After the first step the black liquor has about 20 -

30% solids. At this concentration the rosin soap rises to the surface and is skimmed off. The collected soap is further processed to tall oil. Removal of the soap improves the evaporation operation of the later effects.

The weak black liquor is further evaporated to 65% or even 80% solids ("heavy black liquor") and burned in the recovery boiler to recover the inorganic chemicals for reuse in the pulping process. Higher solids in the concentrated black liquor increases the energy and chemical efficiency of the recovery cycle, but also gives higher viscosity and precipitation of solids (plugging and fouling of equipment). During combustion sodium sulfate is reduced to sodium sulfide by the organic carbon in the mixture:

$$1.\ Na_2SO_4 + 2\ C \rightarrow Na_2S + 2\ CO_2$$

This reaction is similar to thermochemical sulfate reduction in geochemistry.

The molten salts ("smelt") from the recovery boiler are dissolved in a process water known as "weak wash". This process water, also known as "weak white liquor" is composed of all liquors used to wash lime mud and green liquor precipitates. The resulting solution of sodium carbonate and sodium sulfide is known as "green liquor", although it is not known exactly what causes the liquor to be green. This liquid is mixed with calcium oxide, which becomes calcium hydroxide in solution, to regenerate the white liquor used in the pulping process through an equilibrium reaction (Na_2S is shown since it is part of the green liquor, but does not participate in the reaction):

$$2.\ Na_2S + Na_2CO_3 + Ca(OH)_2 \longleftrightarrow Na_2S + 2\ NaOH + CaCO_3$$

Calcium carbonate precipitates from the white liquor and is recovered and heated in a lime kiln where it is converted to calcium oxide (lime).

$$3.\ CaCO_3 \rightarrow CaO + CO_2$$

Calcium oxide (lime) is reacted with water to regenerate the calcium hydroxide used in Reaction 2:

$$4.\ CaO + H_2O \rightarrow Ca(OH)_2$$

The combination of reactions 1 through 4 form a closed cycle with respect to sodium, sulfur and calcium and is the main concept of the so-called recausticizing process wheresodium carbonate is reacted to regenerate sodium hydroxide.

The recovery boiler also generates high pressure steam which is fed to turbogenerators, reducing the steam pressure for the mill use and generating electricity. A modern kraft pulp mill is more than self-sufficient in its electrical generation and normally will provide a net flow of energy which can be used by an associated paper mill or sold to neighboring industries or communities through to the local electrical grid. Additionally, bark and wood residues are often burned in a separate power boiler to generate steam.

Although recovery boilers using G.H. Tomlinson's invention have been in general use since the early 1930s attempts have been made to find a more efficient process for the recovery of cooking chemicals. Weyerhaeuser has operated a Chemrec first generation black liquor entrained flow gasifier successfully at its New Bern plant in North Carolina, while a second generation plant is run in pilot scale at Smurfit Kappa's plant in Piteå, Sweden.

Blowing

The finished cooked wood chips are blown by reducing the pressure to atmospheric pressure. This releases a lot of steam and volatiles. The steam produced can then be used to heat the pulp mill and any excess used in district heating schemes or to drive a steam turbine to generate electrical power. The volatiles are condensed and collected; in the case of northern softwoods this consists mainly of raw turpentine.

Screening

Screening of the pulp after pulping is a process whereby the pulp is separated from large shives, knots, dirt and other debris. The *accept* is the pulp. The material separated from the pulp is called *reject.*

The screening section consists of different types of sieves (screens) and centrifugal cleaning. The sieves are normally set up in a multistage cascade operation because considerable amounts of good fibres can go to the reject stream when trying to achieve maximum purity in the accept flow.

The fiber containing shives and knots are separated from the rest of the reject and reprocessed either in a refiner and/or is sent back to the digester. The content of knots is typically 0.5 - 3.0% of the digester output, while the shives content is about 0.1- 1.0%.

Washing

The brownstock from the blowing goes to the washing stages where the used cooking liquors are separated from the cellulose fibers. Normally a pulp mill has 3-5 washing stages in series. Washing stages are also placed after oxygen delignification and between the bleaching stages as well. Pulp washers use counter current flow between the stages such that the pulp moves in the opposite direction to the flow of washing waters. Several processes are involved: thickening / dilution, displacement and diffusion. Thedilution factor is the measure of the amount of water used in washing compared with the theoretical amount required to displace the liquor from the thickened pulp. Lower dilution factor reduces energy consumption, while higher dilution factor normally gives cleaner pulp. Thorough washing of the pulp reduces the chemical oxygen demand (COD).

Several types of washing equipment are in use:

- Pressure diffusers
- Atmospheric diffusers

- Vacuum drum washers
- Drum displacers
- Wash presses

Bleaching

In a modern mill, brownstock (cellulose fibers containing approximately 5% residual lignin) produced by the pulping is first washed to remove some of the dissolved organic material and then further delignified by a variety of bleaching stages.

In the case of a plant designed to produce pulp to make brown sack paper or linerboard for boxes and packaging, the pulp does not always need to be bleached to a high brightness. Bleaching decreases the mass of pulp produced by about 5%, decreases the strength of the fibers and adds to the cost of manufacture.

Process chemicals

Process chemicals are added to improve the production process:

- Impregnation aids. Surfactants may be used to improve impregnation of the wood chips with the cooking liquors.
- Anthraquinone is used as a digester additive. It works as a redox catalyst by oxidizing cellulose and reducing lignin. This protects the cellulose from degradation and makes the lignin more water-soluble.
- An emulsion breaker can be added in the soap separation to speed up and improve the separation of soap from the used cooking liquors by flocculation.
- Defoamers remove foam and speed up the production process. Drainage of washing equipment is improved and gives cleaner pulp.
- Dispersing agents, detackifiers and complexing agents are keeping the system cleaner and reduce the need for maintenance stops.
- Fixation agents are fixating finely dispersed potential deposits to the fibers and thereby transporting it out of the process.

COMPARISON WITH OTHER PULPING PROCESSES

Pulp produced by the kraft process is stronger than that made by other pulping processes and maintaining a high effective sulfur ratio or sulfidity is important for the highest possible strength. Acidic sulfite processes degrade cellulose more than the kraft process, which leads to weaker fibers. Kraft pulping removes most of the lignin present originally in the wood whereas mechanical pulping processes leave most of the lignin in the fibers. The hydrophobic nature of lignin interferes with the formation of the hydrogen bondsbetween cellulose (and hemicellulose) in the fibers needed for the strength of paper (strength refers to tensile strength and resistance to tearing).

Kraft pulp is darker than other wood pulps, but it can be bleached to make very white pulp. Fully bleached kraft pulp is used to make high quality paper where strength, whiteness and resistance to yellowing are important.

The kraft process can use a wider range of fiber sources than most other pulping processes. All types of wood, including very resinous types like southern pine, and non-wood species like bamboo and kenaf can be used in the kraft process.

BYPRODUCTS AND EMISSIONS

The main byproducts of kraft pulping are crude sulfate turpentine and tall oil soap. The availability of these is strongly dependent on wood species, growth conditions, storage time of logs and chips, and the mill's process. Pines are the most extractive rich woods. The raw turpentine is volatile and is distilled of the digester, while the raw soap is separated from the spent black liquor by decantation of the soap layer formed on top of the liquor storage tanks. From pines the average yield of turpentine is 5–10 kg/t pulp and of crude tall oil is 30–50 kg/t pulp.

Various byproducts containing hydrogen sulfide, methyl mercaptan, dimethyl sulfide, dimethyl disulfide, and other volatile sulfur compounds are the cause of the malodorous air emissions characteristic for pulp mills utilizing the kraft process. Outside the modern mills the odour is perceivable only during disturbance situations, for example when shutting the mill down for maintenance break. This is due to practiced collection and burning of these odorous gases in the recovery boiler along with black liquor. The sulfur dioxideemissions of the kraft pulp mills are much lower than sulfur dioxide emissions from sulfite mills. In modern mills where high dry solids are burned in the recovery boiler hardly anysulfur dioxide leaves the boiler. This is mainly due to higher lower furnace temperature which leads to higher sodium release from the black liquor droplets that can react with sulfur dioxide forming sodium sulfate.

Pulp mills are almost always located near large bodies of water due to their former substantial demands. Delignification of chemical pulps released considerable amounts of organic material into the environment, particularly into rivers or lakes.

The wastewater effluent can also be a major source of pollution, containing lignins from the trees, highbiological oxygen demand (BOD) and dissolved organic carbon (DOC), along with alcohols, chlorates, heavy metals, and chelating agents.

Reducing the environmental impact of this effluent is accomplished by closing the loop and recycling the effluent where possible, as well as employing less damaging agents in the pulping and bleaching processes. The process effluents are treated in a biological effluent treatment plant, which guarantees that the effluents are not toxic in the recipient.

SULFITE PROCESS

The sulfite process produces wood pulp which is almost pure cellulose fibers by using various salts of sulfurous acid to extract the lignin from wood chips in large pressure vessels called digesters. The salts used in the pulping process are either sulfites (SO_3^{2-}), or bisulfites (HSO_3^-), depending on the pH. The counter ion can be sodium (Na^+),calcium (Ca^{2+}), potassium (K^+), magnesium (Mg^{2+}) or ammonium (NH_4^+).

HISTORY

The use of wood to make pulp for paper began with the development of mechanical pulping in Germany by F.G. Keller in the 1840s. Chemical processes quickly followed, first with J. Roth's use of sulfurous acid to treat wood, followed by Benjamin Chew Tilghman's US patent on the use of calcium bisulfite, $Ca(HSO_3)_2$, to pulp wood in 1867. Almost a decade later in 1874 the first commercial sulfite process|sulfite pulp mill was built in Sweden. It used magnesium as the counter ion and was based on work by Carl Daniel Ekman.

By 1900 sulfite pulping had become the dominant means of producing wood pulp, surpassing mechanical pulping methods. The competing chemical pulping process, the sulfate or kraft process was developed by Carl F. Dahl in 1879 and the first kraft mill started (in Sweden) in 1890. The first sulphite mill in the United States was the Richmond Paper Company in Rumford, Rhode Island in the mid-1880s. The invention of the recovery boiler by G.H. Tomlinson in the early 1930s allowed kraft mills to recycle almost all of their pulping chemicals. This, along with the ability of the kraft process to accept a wider variety of types of wood and produce stronger fibers made the kraft process the dominant pulping process starting in the 1940s. Sulfite pulps now account for less than 10% of the total chemical pulp production and the number of sulfite mills continues to decrease.

Magnesium was the standard counter ion until calcium replaced it in the 1950s.

PROCESSES INVOLVED IN SULFITE PULPING

Pulping liquor preparation

The pulping liquor for most sulfite mills is made by burning sulfur with the correct amount of oxygen to give sulfur dioxide, which is then absorbed into water to give sulfurous acid.

$$S + O_2 \rightarrow SO_2$$

$$SO_2 + H_2O \rightleftharpoons H_2SO_3$$

Care must be taken to avoid the formation of sulfur trioxide since it gives undesired sulfuric acid when it is dissolved in water.

$$2\,SO_2 + O_2 \rightarrow 2SO_3$$
$$SO_3 + H_2O \rightleftharpoons H_2SO_4$$

Sulfuric acid is undesirable since it promotes hydrolysis of cellulose without contributing to delignification.

The cooking liquor is prepared by adding the counter ions as hydroxides or carbonates. The relative amounts of each species present in the liquid depend largely on the relative amounts of sulfurous used. For monovalent (Na^+, K^+ and NH^{4+}) hydroxides, MOH:

$$H_2SO_3 + MOH \rightarrow MHSO_3 + H_2O$$
$$MHSO_3 + MOH \rightarrow M_2SO_3 + H_2O$$

For divalent (Ca^{2+}, Mg^{2+}) carbonates, MCO_3:

$$MCO_3 + 2H_2SO_3 \rightarrow M(HSO_3)_2 + CO_2 + H_2O$$
$$M(HSO_3)_2 + MCO_3 \rightarrow 2\,MSO_3 + CO_2 + H_2O$$

Pulping

Sulfite pulping is carried out between pH 1.5 and 5, depending on the counterion to sulfite (bisulfite) and the ratio of base to sulfurous acid. The pulp is in contact with the pulping chemicals for 4 to 14 hours and at temperatures ranging from 130 to 160 °C (266 to 320 °F), again depending on the chemicals used.

Most of the intermediates involved in delignification in sulfite pulping are resonance-stabilized carbocations formed either by protonation of carbon-carbon double bonds or acidic cleavage of ether bonds which connect many of the constituents of lignin. It is the latter reaction which is responsible for most lignin degradation in the sulfite process. Theelectrophilic carbocations react with bisulfite ions (HSO_3^{-})to give sulfonates.

$$R\text{-}O\text{-}R' + H^+ \rightarrow R^+ + R'OH$$
$$R^+ + HSO_3^{-} \rightarrow R\text{-}SO_3H$$

The sulfite process does not degrade lignin to the same extent that the kraft process does and the lignosulfonates from the sulfite process are useful byproducts.

Chemical recovery

The spent cooking liquor from sulfite pulping is usually called brown liquor, but the terms red liquor, thick liquor and sulfite liquor are also used (compared to black liquor in thekraft process).

Pulp washers, using countercurrent flow, remove the spent cooking chemicals and degraded lignin and hemicellulose. The extracted brown liquor is concentrated, in multiple effect evaporators.

The concentrated brown liquor can be burned in the recovery boiler to generate steam and recover the inorganic chemicals for reuse in the pulping process or it can be neutralized to recover the useful byproducts of pulping.

Recent developments in Chemrec's black liquor gasification process, adapting the technology to use in the sulfite pulping process, could make second generation biofuels production an alternative to the conventional recovery boiler technology.

The sulfite process can use calcium, ammonium, magnesium or sodium as a base.

Calcium-based

The earliest process used calcium, obtained as inexpensive calcium carbonate and there was little incentive to recover the inorganic materials.

Ammonia-based

Ammonia-based processes do not allow recovery of the pulping chemicals since ammonia or ammonium salts are oxidized to nitrogen and nitrogen oxides when burned.

Magnesium-based

The recovery process used in magnesium-based sulfite pulping the "Magnefite" process is well developed. The concentrated brown liquor is burned in a recovery boiler, producing magnesium oxide and sulfur dioxide, both of which are recovered from the flue gases. Magnesium oxide is recovered in a wet scrubber to give a slurry of magnesium hydroxide.

$$MgO + H_2O \rightarrow Mg(OH)_2$$

This magnesium hydroxide slurry is then used in another scrubber to absorb sulfur dioxide from the flue gases producing a magnesium bisulfite solution that is clarified, filtered and used as the pulping liquor.

$$Mg(OH)_2 + 2\ SO_2 \rightarrow Mg(HSO_3)_2$$

Sodium-based

Sodium-based processes use a recovery system similar to that used in the kraft recovery process, except that there is no "lime cycle".

APPLICATIONS

The sulfite process is acidic and one of the drawbacks is that the acidic conditions hydrolyze some of the cellulose, which means that sulfite pulp fibers are not as strong as kraft pulp fibers. The yield of pulp (based on wood used) is higher than for kraft pulping and sulfite pulp is easier to bleach.

Commodity

Sulfite pulp remains an important commodity, especially for specialty papers and as a source of cellulose for non-paper applications. It is used to make fine paper, tissue,glassine. and to add strength to newsprint.

Dissolving pulp

A special grade of bleached sulfite pulp is known as *dissolving pulp* which is the raw material for a wide variety of cellulose derivatives, for example rayon, cellophane,cellulose acetate and methylcellulose.

Rayon is a reconstituted cellulose fiber used to make many fabrics.

Cellophane is a clear reconstituted cellulose film used in wrapping and windows in envelopes.

Cellulose acetate was used to make flexible films for photographic use, computer tapes and so on and also to make fibers.

Methylcellulose and other cellulose ether derivatives are used in a wide range of everyday products from adhesives to baked goods to pharmaceuticals.

BYPRODUCTS

Sulfite pulping is generally less destructive than kraft pulping, so there are more usable byproducts.

Lignosulfonates

Chief among sulfite process byproducts are lignosulfonates, which find a wide variety of uses whereas relatively inexpensive agent is needed to make a water dispersion of a water-insoluble material. Lignosulfonates are used in tanning leather, making concrete, drilling mud, drywall and so on.

Oxidation of lignosulfonates was used to produce vanillin (artificial vanilla), and this process is still used by one supplier (Borregaard, Norway) while all North American production by this route ceased in the 1990s.

Other byproducts

Acid hydrolysis of cellulose during sulfite pulping produces monosaccharides, predominantly mannose, which can be fermented to produce ethanol.

SODA PULPING

Soda pulping is a chemical process for making wood pulp with sodium hydroxide as the cooking chemical. In the *Soda-AQ* process, anthraquinone (AQ) may be used as a pulping additive to decrease the carbohydrate degradation. The soda process gives pulp with lower tear strength than other chemical pulping processes (sulfite process andkraft process), but has still limited use for easy pulped materials like straw and some hardwoods.

HISTORY

Soda pulping was one of the first chemical pulping methods and was invented in 1851 by Burgess (USA) and Watts (England). The first mill was started in 1860 in the USA. In 1865 they patented a method for recovery of the cooking liquors by incineration of the spent liquor. Many of the early soda mills converted to kraft mills once it was discovered.

PRODUCTION

Around 5%-10% of paper production worldwide is produced from agricultural crops, valuing agricultural paper production at between $5 billion and $10 billion. The most notable of these agricultural crops are wheat straw and bagasse. Using agricultural crops rather than wood has the added advantage of reducing deforestation.

Due to the ease with which bagasse can be chemically pulped, bagasse requires less bleaching chemicals than wood pulp to achieve a bright, white sheet of paper.

Most chemical bagasse pulp mills concentrate the spent reaction chemicals and combust them to power the paper-mills and to recover the reaction chemicals.

Soda pulping as solution for silicate scaling

Many grasses, bagasse, bamboo and some tropical hardwoods contain much silicates that may cause sodium aluminum silicate scales. Moderate amounts of silicates can be controlled with purging lime mud or lime kiln ash. Silicate removal from green liquor in a soda mill can be achieved by lowering the pH of the liquor with CO_2-containing flue gases from the lime kiln or other sources. No commercial silicate removal system is available for the kraft process, but it can handle the small amounts of silicates from northern woods.

GRAMMAGE

Basis weight and grammage are terms used in the pulp and paper industry and also for fabric industry to denote a measure of mass of the product per unit of area for a type of fabric, paper or paperboard. Two ways of expressing grammage are commonly used:

- Expressed in grams per square meter (g/m^2), paper density is also known as *grammage*. This is the measure used in most parts of the world.
- Expressed in terms of the mass (expressed as weight) per number of sheets, it is known as *basis weight*. The convention used in the United States and a few other countries using US paper sizes is pounds of a ream of 500 (or in some cases 1000) sheets of a given (raw, still uncut) basis size. Japanese paper is expressed as the weight in kg of 1,000 sheets.

GRAMMAGE

In the metric system, the mass per unit area of all types of paper and paperboard is expressed in terms of grams per square meter (g/m^2). This quantity is commonly called*grammage* in both English and French (ISO 536), though printers in most English-speaking countries still refer to the "weight" of paper.

Typical office paper has 80 g/m^2, therefore a typical A4 sheet ($^1D_{16}$ m^2) weighs 5 g. The unofficial unit symbol "gsm" instead of the standard "g/m^2" is also widely encountered in English speaking countries.

Typically grammage is measured in paper mill on-line by Quality Control System (QCS) and verified by laboratory measurement.

BASIS WEIGHT

In countries that use US paper sizes, a less direct measure known as *basis weight* is used in addition to or instead of grammage. The basis weight of paper is the density of paper expressed in terms of the mass of a ream of given dimensions and a sheet count. In the US system, the weight is specified in avoirdupois pounds and the sheet count of apaper ream is usually 500 sheets. However, the mass specified is not the mass of the ream that is sold to the customer. Instead, it is the mass of the uncut "basis ream" in which the sheets have some larger size. Often, this is a size used during the manufacturing process before the paper was cut to the dimensions in which it is sold. So, to compute the mass per area, one must know

- The mass of the basis ream,
- The number of sheets in that ream, and
- The dimensions of an "uncut" sheet in that ream.

The standard dimensions and sheet count of a ream vary according to the type of paper. These "uncut" basis sizes are not normally labelled on the product, are not formally standardized, and therefore have to be guessed or inferred somehow from trading practice.

Historically, this convention is the product of pragmatic considerations such as the size of a sheet mold.

By using the same basis sheet size for the same type of paper, consumers can easily compare papers of differing brands. Twenty pound bond paper is always lighter and thinner than 32 pound bond, no matter what its cut size. And 20 pound bond *letter size* and 20 pound bond *legal size* papers are the same weight paper having different cut size.

However, a sheet of common copy paper that has a basis weight of 20 pounds (9.1 kg) does not have the same mass as the same size sheet of coarse paper (newsprint).

In the former case, the standard ream is 500 sheets of 17-by-22-inch (432 by 559 mm) paper, and in the latter, 500 sheets of 24-by-36-inch (610 by 914 mm) paper. Here are some basic ream sizes for various types of paper. Units are inches except where noted.

Paper Type	Paper Size	- Number of sheets
Bond, writing, ledger	17 × 22	- 500 sheets
Manuscript cover	18 × 31	- 500 sheets
Blotting	19 × 24	- 500 sheets
Box cover	20 × 24	- 500 sheets
Cover	20 × 26	- 500 or 1000 sheets
Bristol and tag	22½ × 28½	- 500 sheets
Tissue	24 × 36	- 480 sheets
Newsprint	24 × 36	- 500 sheets
Hanging, waxing, bag, etc.	24 × 36	- 500 sheets
Book, Text, Offset	25 × 38	- 500 sheets
Index bristol	25½ × 30½	- 500 sheets
Paperboard (all types)	12 × 12	- 1000 sheets (1,000 square feet per ream)

Sheets 17 by 22 inches (432 by 559 mm) can be cut into four 8.5-by-11-inch (216 by 279 mm) sheets, a standard for business stationery known conventionally as *letter sized paper*. So, the 17-by-22-inch (432 by 559 mm) ream became commonly used. The 25-by-38-inch (635 by 965 mm) book-paper ream developed because such a size can easily be cut into sixteen 6 by 9-inch (229 mm) book sized sheets without significant waste.

Early newsprint presses printed sheets 2 by 3 feet (0.610 by 0.914 m) in size, and so the ream dimensions for newsprint became 24 by 36 inches (610 by 914 mm), with 500 sheets to a ream. Newsprint was made from ground wood pulp, and ground wood hanging paper (wallpaper) was made on newsprint machines. Newsprint was used as wrapping paper, and the first paper bags were made from newsprint. The newsprint ream standard also became the standard for packaging papers, even though in packaging papers kraft pulp rather than ground wood was used for greater strength.

Paper weight is sometimes stated using the "#" symbol. For example, "20#" means "20 pounds per basis ream of 500 sheets". When the density of a ream of paper is given in pounds, it is often accompanied by its "M weight". The M weight is the weight (in pounds) of 1000 cut sheets. Paper suppliers will often charge by M weight, since it is always consistent within a specific paper size, and because it allows a simple weight calculation for shipping charges.

For example, a 500-sheet ream of 20# 8.5-by-11-inch (216 by 279 mm) copy paper may be specified "10 M". 1000 cut sheets (or two reams) will weigh 10 lb (4.5 kg), half of the four reams of cut paper resulting from the 20# basis ream of 17-by-22-inch (432 by 559 mm) paper.

CALIPER

Paper thickness, or caliper, is a common measurement specified and required for certain printing applications. Since a paper's density is typically not directly known or specified, the thickness of any sheet of paper cannot be calculated by any method. Instead, it is measured and specified separately as its caliper. However, paper thickness for most typical business papers might be similar across comparable brands. If thickness is not specified for a paper in question, it must be either measured or guessed based on a comparable paper's specification.

Caliper is usually measured in micrometres (μm), or in the US also in mils (1 mil = 0.001 in = 25.4 μm). Commonly, 20-lb bond ranges between roughly 0.0038 to 0.0045 inches (97 to 114 μm) in thickness.

WOODCHIPS

Fig. Large woodchipper (Europe Chippers model C1175). This type of machine is used to chip large pieces of wood

Woodchips are a medium-sized solid material made by cutting, or chipping, larger pieces of wood.

Woodchips may be used as a biomass solid fuel and are raw material for producing wood pulp. They may also be used as an organicmulch in gardening, landscaping, restoration ecology, bioreactors for denitrification and mushroom cultivation. According to the different chemical and mechanical properties of the masses, the wood logs are mostly peeled, and the bark chips and the wood chips are processed in different processes. The process of making wood chips is called woodchipping and is done with a woodchipper.

RAW MATERIALS

The raw materials of wood chips can be pulpwood, wood plantations, waste wood and residuals from construction, agriculture, landscaping, logging, and sawmills and locally grown and harvested fuel crops.

PRODUCTION

A woodchipper is a machine used for reducing wood to smaller pieces. There are several types of woodchippers depending of the further processing of the woodchips. For industrial use, the woodchippers are large, stationary installations.

Pulp and paper industry

Wood chips used for chemical pulp must be relatively uniform in size and free of bark. The optimum size varies with the wood species. It is important to avoid damage to the wood fibres as this is important for the pulp properties. For roundwood it is most common to use disk chippers. A typical size of the disk is 2.0 - 3.5 m in diameter, 10 – 25 cm in thickness and weight is up to 30 tons. The disk is fitted with 4 to 16 knives and driven with motors of ½ - 2 MW. Drum chippers are normally used for wood residuals from saw mills or other wood industry.

Methods of conveyance

There are four potential methods to move woodchips: pneumatic, conveyor belt, hopper with direct chute, batch system (manual conveyance).

APPLICATIONS

Woodchips are used primarily as a raw material for technical wood processing. In industry, processing of bark chips is often separated after peeling the logs due to different chemical properties.

Wood pulp

Only the heartwood and sapwood are useful for making pulp. Bark contains relatively few useful fibres and is removed and used as fuel to provide steam for use in the pulp mill. Most pulping processes require that the wood be chipped and screened to provide uniform sized chips.

Mulch

Woodchipping is also used to produce landscape and garden woodchips mulch. It is used for water conservation, weed control, reducing and preventing soil erosion, and for supporting germination of native seeds and acorns in habitat revegetation-ecological restoration projects. As the ramial chipped wood decompose it improve the soil structure, permeability, bioactivity, and nutrient availability. Woodchips when used as a mulch are at least three inches thick.

Playground surfacing

Wood chips can be reprocessed into an extremely effective playground surfacing material, or impact-attenuation surface. When used as playground surfacing (soft fall, cushion fall, or play chip, as it is sometimes known), woodchips can be very effective in lessening the impact of falls from playground equipment. When spread to depths of one foot (30 centimeters) playground wood chips can be effective at reducing impacts in falls up to 11 feet (3 meters). Playground woodchip is also an environmentally friendly alternative to rubber type playground surfaces.

Bioreactor

Woodchip piles at the edge of a field can inhibit nitrates from running off into water tiles. They are a simple measure for farmers to reduce nitrate pollution of the watershed without them having to change their land management practice. A 2011 study showed that most of the nitrate removal was due to heterotrophic denitrification. A 2013 experiment from Ireland showed that after 70 days of startup, a woodchip pile loaded with liquid pig manure at 5 $L/m^2/day$ removed an average of 90% of nitrate in the form ofammonium after one month. A January 2015 study from Ohio State University showed very low nitrogen gas, i.e. greenhouse gas emissions from nitrate transformation under the anaerobic conditions of the wood chip bioreactor. Scientists constructed a model for water flow and nitrate removal kinetics which can be used to design denitrificationbeds. It is unknown if other nutrients like phosphorus or pathogens are affected by the bioreactor as well.

Fuel

Woodchips have been traditionally used as solid fuel for space heating or in energy plants to generate electric power from renewable energy. The main source of forest chips in Europe and in most of the countries have been logging residues. It is expected that the shares of stumps and roundwood will increase in the future. As of 2013 in the EU, the estimates for biomass potential for energy, available under current conditions including sustainable use of the forest as well as providing wood to the traditional forest sectors, are: 277 million m^3, for above ground biomass and 585 million m^3 for total biomass.

The newer fuel systems for heating use either woodchips or wood pellets. The advantage of woodchips is cost, the advantage of wood pellets is the controlled fuel value. The use of woodchips in automated heating systems, is based on a robust technology.

The size of the woodchips is particularly important when burning woodchip in small plants. Unfortunately there are not many standards to decide the fractions of woodchip. One standard is the GF60 which is commonly used in smaller plants, including small industries, villas, and apartment buildings. "GF60" is known as "Fine, dry, small chips". The requirements for GF60 are that the moisture is between 10–30% and the fractions of the woodchips are distributed as follows: 0–3.5mm: <8%, 3.5–30mm: <7%, 30–60mm: 80–100%, 60–100mm: <3%, 100–120mm: <2%.

The energy content in one cubic metre is normally higher than in one cubic metre wood logs, but can vary greatly depending on moisture. The moisture is decided by the handling of the raw material. If the trees is taken down in the winter and is left to dry under the summer, with tears in the bark and covered so rain can't reach to them and then is chipped in the fall the woodchip will get an moisture of approx. 20–25%. The energy content is then approx 3.5–4.5kWh/kg (~150–250 kg/cubic metre).

Coal power plants have been converted to run on woodchips, which is fairly straightforward to do, since they both use an identical steam turbine heat engine, and the cost of woodchip fuel is comparable to coal.

Solid biomass is an attractive fuel for addressing the concerns of the energy crisis and climate change, since the fuel is affordable, widely available, close to carbon neutral and thus climate-neutral in terms of carbon dioxide (CO2), since in the ideal case only the carbondioxide which was drawn in during the tree's growth and stored in the wood is released into the atmosphere again. It is sustainable as long as crops are allowed to regrow; In most cases, biomass is not carbon neutral as wood is not regrown and the efficiency of biomass operations produce more pollutants than the processes they replace.

Waste and emissions

Compared to the solid waste disposal problems of coal and nuclear fuels, woodchip fuel's waste disposal problems are less grave; in a study from 2001 fly ash from wood chip combustion had 28.6 mg cadmium/kg dry matter. Compared to fly ash from burning of straw, cadmium was bound more heavily, with only small amounts of cadmium leached. It was speciated as a form of cadmium oxide, cadmium silicate ($CdSiO_3$); authors noted that adding it to agricultural or forest soils in the long-term could cause a problem with accumulation of cadmium.

Like coal, wood combustion is a known source of mercury emissions, particularly in northern climates during winter.

The mercury is both gaseous as elemental mercury (especially when wood pellets are burned) or mercury oxide, and solid PM2.5 particulate matter when untreated wood is used.

When wood burning is used for space heating, indoor emissions of 1,3-butadiene, benzene, formaldehyde and acetaldehyde, which are suspected or known carcinogenic compounds, are elevated. The cancer risk from these after exposure to wood smoke is estimated to be low in developed countries .

Certain techniques for burning woodchips result in the production of biochar - effectively charcoal - which can be either utilised as charcoal, or returned to the soil, since wood ash can be used as a mineral-rich plant fertilizer. The latter method can result in an effectively carbon-negative system, as well as acting as a very effective soil conditioner, enhancing water and nutrient retention in poor soils.

Automated handling of solid fuel

Unlike the smooth, uniform shape of manufactured wood pellets, woodchip sizes vary and are often mixed with twigs and sawdust. This mixture has a higher probability of jamming in small feed mechanisms. Thus, sooner or later, one or more jams is likely to occur. This reduces the reliability of the system, as well as increasing maintenance costs. Despite what some pellet stove manufacturers may say, researchers who are experienced with woodchips, say they are not compatible with the 2 inch (5 cm) auger used in pellet stoves.

Micro combined heat and power

Wood is occasionally used to power engines, such as steam engines, Stirling engines, and Otto engines running on woodgas. As of 2008, these systems are rare, but as technology and the need for it develops, it is likely to be more common in the future. For the time being, wood can be increasingly used for heating applications. This will reduce the demand for heating oil, and thereby allow a greater percentage of fuel oil to be used for applications such as internal combustion engines, which are less compatible with wood based fuel and other solid biomass fuels. Heating applications generally do not require refined or processed fuels, which are almost always more expensive.

Comparison to other fuels

Woodchips are similar to wood pellets, in that the movement and handling is more amenable to automation than cord wood, particularly for smaller systems. Woodchips are less expensive than wood pellets and are theoretically more energy efficient than pellets, because less energy is required for manufacturing, processing, and transportation; however, this assumes that they are consumed in an appropriately designed burner, and as of 2008, these are mostly only available in large systems designed for commercial or institutional use, which have been very successful in terms of performance, cost, reliability, and efficiency. Woodchips are less expensive than cord wood, because the harvesting is faster and more highly automated. Woodchips are of greater supply, partly because all parts of a tree can be chipped, whereas small limbs and branches can require much labor convert to cord wood. Cordwood generally needs to be "seasoned" or "dry" before it can be burned cleanly and efficiently. On the other hand, woodchip systems are typically designed to cleanly and efficiently burn "green chips" with very high moisture content of 43–47% (wet basis).

ENVIRONMENTAL ASPECTS

If woodchips are harvested through sustainable forestry practices, then this is considered a source of renewable energy. On the other hand, harvesting practices, such asclearcutting large areas, are highly damaging to forest ecosystems.

Theoretically, whole-tree chip harvesting does not have as high a solar energy efficiency compared to short rotation coppice; however, it can be an energy-efficient and low-cost method of harvesting. In some cases, this practice may be controversial when whole-tree harvesting may often be associated with clear cutting and perhaps other questionable forestry practices.

Waste processing

Woodchips, and bark chips, can be used as bulking agents in industrial composting of municipal biodegradeable waste, particularly biosolids.

Woodchip biomass does not have the waste disposal issues of coal and nuclear power, since wood ash can be used directly as a mineral-rich plant fertilizer.

Forest fire prevention

Woodchip harvesting can be used in concert with creating man made firebreaks, which are used as barriers to the spread of wildfire. Undergrowth coppice is ideal for chipping, and larger trees may be left in place to shade the forest floor and reduce the rate of fuel accumulation.

MARKET PRODUCTS, SUPPLY AND DEMAND

Currently, domestic or residential sized systems are not available in products for sale on the general market. Homemade devices have been produced, that are small-scale, clean-burning, and efficient for woodchip fuels. Much of the research activity to date, has consisted of small budget projects that are self-funded. The majority of funding for energy research has been for liquid biofuels.

United States

"Wood chip costs usually depend on such factors as the distance from the point of delivery, the type of material (such as bark, sawmill residue or whole-tree chips), demand by other markets and how the wood fuel is transported. Chips delivered directly to the (powerplant) station by truck are less expensive than those delivered ... and shipped by railcar. The range of prices is typically between US$18 to US$30 per (wet)-ton delivered."

In 2006, prices were US$15 and US$30 per wet-ton in the northeast.

In the 20 years leading up to 2008, prices have fluctuated between US$60–70/oven-dry metric ton (odmt) in the southern states, and between US$60/odmt and US$160/odmt in the Northwest.

Europe

In several well wooded European countries (e.g. Austria, Finland, Germany, Sweden) wood chips are becoming an alternative fuel for family homes and larger buildings due to the abundant availability of wood chips, which result in low fuel costs. The European Union is promoting wood chips for energy production in the EU Forest action plan 2007-2011. The total long term potential of wood chips in the EU it is estimated to be 913 million m3.

Fig. Large woodchipper in Germany

Japan

After a long period of negative scores, the demand of wood chip for paper manufacturing started increasing again. Starting in the last quarter of 2013, orders for printing paper and card board increased before the consumption tax increase then by weakening yen, import of papers like copy paper decreases and export of paper increases, which stimulate paper production in Japan. Softwood chip prices from the United States increased by 12% compared to October 2013 and softwood chip prices from Australia increased by 7%.

DISSOLVING PULP

Dissolving pulp, also called dissolving cellulose, is a bleached wood pulp or cotton linters that has a high cellulose content (> 90%). It has special properties including as a high level of brightness and uniform molecular-weight distribution.

This pulp is manufactured for uses that require a high chemical purity, and particularly low hemicellulosecontent, since the chemically similar hemicellulose can interfere with subsequent processes. Dissolving pulp is so named because it is not made into paper, but dissolved either in a solvent or by derivatization into a homogeneous solution, which makes it completely chemically accessible and removes any remaining fibrous structure. Once dissolved, it can be spun into textile fibers (viscose or Lyocell), or chemically reacted to produce derivatized celluloses, such cellulose triacetate, a plastic-like material formed into fibers or films, or cellulose ethers such as methyl cellulose, used as a thickener.

MANUFACTURE

Pulpwood

Dissolving pulp is mainly produced chemically from the pulpwood in a process that has a low yield (30 - 35% of the wood). This makes up of about 85 - 88% of the production.Dissolving pulp is made from the sulfite process or the kraft process with an acid prehydrolysis step to remove hemicelluloses. For the highest quality, it should be derived from fast-grown hardwoods with low non-cellulose content.

The sulfite process produces pulp with a cellulose content up to 92 percent. It can use ammonium, calcium, magnesium or sodium as a base. The prehydrolysis sulfate processproduces pulp with a cellulose content up to 96%.

Special alkaline purification treatments can yield even higher cellulose levels: up to 96 percent for the sulfite process and up to 98 percent for the sulfate process.

Cotton linters

The minor part is produced from the shortest cotton linters, normally second cut. These are washed mechanically and chemically to remove proteins, waxes, pectins and otherpolysaccharides. This is bleached to get the required brightness. Dissolving pulp from cellulose linters gives the purest cellulose and is used for manufacture acetate plastics and high-viscosity cellulose ethers.

APPLICATIONS

Dissolving pulp is used in production of regenerated cellulose. In the regenerated cellulose process the cellulose is converted to cellulose xanthate which dissolves easily in caustic soda. The resulting viscous liquid can be extruded through spinnerettes and regenerated as man-made fibres. Cellulose can also be dissolved in some organic solvents directly and processed to regenerate the cellulose fibres in different forms. The lyocell process uses an amine oxide to dissolve cellulose and Tencel is the only commercial example of this direct-dissolution process, which unlike the viscose process is pollution-free.

The 90-92% cellulose content sulfite pulps are used mostly to make textiles (like rayon) and cellophane. The 96-% cellulose content sulfate pulps are used to make rayon yarn for industrial products such as tire cord, rayon staple for high-quality fabrics, and various acetate and other specialty products.

As a raw material of cellulose derivatives, dissolving pulp is used in carboxymethyl cellulose (CMC), methyl cellulose (MC), hydroxypropyl cellulose

(HPC), hydroxyethyl cellulose(HEC), etc. Since dissolving pulp is highly refined, it is a product of high whiteness with few impurities making it suitable in specialty paper-related products such as filter paper andvulcanized fibre.

Cellulose powder is dissolving pulp that has undergone acid hydrolysis, been mechanically disintegrated and made into fine powder.

This pulp is used as a filler for urea-formaldehyde resins and melamine resin products.

FLUFF PULP

Fluff pulp (also called *comminution pulp* or *fluffy pulp*) is a type of chemical pulp made from long fibre softwoods. Important parameters for fluff pulp are bulk and water absorbency.

HISTORY

Fluff pulp was first developed for use in disposable sanitary napkins. Kotex first advertisement for products made with wood pulp (Cellucotton) appeared in 1921.

Disposable diaper producers also were early to convert to fluff pulp because of its low cost and high absorbency. Normal usage of fluff pulp in a diaper was about 55 grams.

In the 1980s started the commercialization of air-laid paper, which gave better bulk, porosity, strength, softness and water absorption properties compared with normal tissue paper.

Also in the 1980s started the use of superabsorbents in diapers and reduced the need for fluff pulp and is now down to 15 grams or even less. The demand of the pulp in diapers has gone from being the absorbent of liquid to giving the products dry and wet strength.

MANUFACTURE

More than 90% of the fluff pulps are fully bleached chemical softwood pulps, of which more than 90% are kraft pulps.

The most common raw material source for fluff pulps are southern bleached softwood kraft (SBSK) from slash pine. SBSK from other species and NBSK are also used to make fluff pulp. Thicker fibres are preferred to improve the bulk.

Fluff pulp is normally made rolls (reels) on a drying machine (a simplified Fourdrinier machine).

The objective of the drying / sheeting operation is to produce a uniform sheet (paper density, moisture and strength) to the converting operation.

The pulp may be impregnated with debonders before drying to ease defibration. The worldwide production of fluff pulps amounts to about 3.5 million tons.

APPLICATIONS

Fluff pulps are used as raw material in the absorbent core of personal care products such as diapers, feminine hygiene products, air-laid absorbent towelling, as such or with superabsorbents and/or synthetical fibres. More than 80% of the pulps are used in baby diapers.

The most demanding application of fluff pulps is in air-laid products, used in serving utensils, various towel applications in homes, in the industry and in hospitals. Fluff pulp for air-laid products are defibrized in a hammermill. Defibration is the process of freeing the fibres from each other before entering the papermachine. Important parameters for dry defibration are shredding energy and knot content.

BLEACHING OF WOOD PULP

Bleaching of wood pulp is the chemical processing carried out on various types of wood pulp to decrease the color of the pulp, so that it becomes whiter. The main use of wood pulp is to make paper where whiteness (similar to but not exactly the same as "brightness") is an important characteristic.

PAPER BRIGHTNESS

Brightness is a measure of how much light is reflected by paper under specified conditions and is usually reported as a percentage of how much light is reflected, so a higher number represents a brighter or whiter paper. In the US, the TAPPI T 452 or T 525 standards are used. The international community uses ISO standards. The following table shows how the two systems rate high brightness papers, but there is no simple way to convert between the two systems because the test methods are so different. Note that the ISO rating is higher and can go above 100. This is because today's white paper manufacturing uses fluorescent whitening agents (FWA). Because the ISO standard only measures a narrow range of blue light, it is not an adequate measure for the actual whiteness or brightness.

TAPPI Brightness	ISO Brightness
84	88
92	104
96	108
97	109+

Newsprint ranges from 55-75 ISO brightness. Writing and printer paper would typically be as bright as 104 ISO.

While the results are the same, the processes and fundamental chemistry involved in bleaching chemical pulps (like kraft or sulfite) are very different from those involved in bleaching mechanical pulps (like stoneground, thermomechanical or chemithermomechanical). Chemical pulps contain very little lignin while mechanical pulps contain most of the lignin that was present in the wood used to make the pulp. Lignin is the main source of color in pulp due to the presence of a variety of chromophores naturally present in the wood or created in the pulp mill.

BLEACHING MECHANICAL PULPS

Mechanical pulp retains most of the lignin present in the wood used to make the pulp and thus contain almost as much lignin as they do cellulose and hemicellulose. It would be impractical to remove this much lignin by bleaching, and undesirable since one of the big advantages of mechanical pulp is the high yield of pulp based on wood used.

Therefore the objective of bleaching mechanical pulp (also referred to as brightening) is to remove only the chromophores (color-causing groups). This is possible because the structures responsible for color are also more susceptible to oxidation or reduction.

Alkaline hydrogen peroxide is the most commonly used bleaching agent for mechanical pulp. The amount of base such as sodium hydroxide is less than that used in bleaching chemical pulps and the temperatures are lower. These conditions allow alkaline peroxide to selectively oxidize non-aromatic conjugated groups responsible for absorbing visible light. The decomposition of hydrogen peroxide is catalyzed by transition metals, and iron, manganese and copper are of particular importance in pulp bleaching. The use ofchelating agents like EDTA to remove some of these metal ions from the pulp prior to adding peroxide allows the peroxide to be used more efficiently. Magnesium salts andsodium silicate are also added to improve bleaching with alkaline peroxide.

Sodium dithionite ($Na_2S_2O_4$), also known as sodium hydrosulfite, is the other main reagent used to brighten mechanical pulps. In contrast to hydrogen peroxide, which oxidizes the chromophores, dithionite reduces these color-causing groups. Dithionite reacts with oxygen, so efficient use of dithionite requires that oxygen exposure be minimized during its use.

Chelating agents can contribute to brightness gain by sequestering iron ions, for example as EDTA complexes, which are less colored than the complexes formed between iron and lignin.

The brightness gains achieved in bleaching mechanical pulps are temporary since almost all of the lignin present in the wood is still present in the pulp. Exposure to air and light can produce new chromophores from this residual lignin. This is why newspaper yellows as it ages. yellowing also occurs due to the acidic sizing

BLEACHING OF RECYCLED PULP

Hydrogen peroxide and sodium dithionite are used to increase the brightess of deinked pulp. The bleaching methods are similar for mechanical pulp in which the goal is to make the fibers brighter.

BLEACHING CHEMICAL PULPS

Chemical pulps, such as those from the kraft process or sulfite pulping, contain much less lignin than mechanical pulps, (<5% compared to approximately 40%). The goal in bleaching chemical pulps is to remove essentially all of the residual lignin, hence the process is often referred to as delignification. Sodium hypochlorite (household bleach) was initially used to bleach chemical pulps, but was largely replaced in the 1930s by chlorine. Concerns about the release of organochlorine compounds into the environment prompted the development of Elemental Chlorine Free (ECF) and Totally Chlorine Free (TCF) bleaching processes.

Delignification of chemical pulps is rarely a single step process and is frequently composed of four or more discrete steps. These steps are given a letter designation, and these are given in the following table:

Chemical or process used	Letter designation
Chlorine	C
Sodium hypochlorite	H
Chlorine dioxide	D
Extraction with sodium hydroxide	E
Oxygen	O
Alkaline hydrogen peroxide	P
Ozone	Z
Chelation to remove metals	Q
Enzymes (especially xylanase)	X
Peracids (peroxy acids)	Paa
Sodium dithionite (sodium hydrosulfite)	Y

A bleaching sequence from the 1950s could look like: CEHEH . The pulp would have been exposed to chlorine, extracted (washed) with a sodium hydroxide solution to remove lignin fragmented by the chlorination, treated with sodium hypochlorite, washed with sodium hydroxide again and given a final treatment with hypochlorite. An example of a modern totally chlorine-free (TCF) sequence is OZEPY where the pulp would be treated with oxygen, then ozone, washed with sodium hydroxide then treated in sequence with alkaline peroxide and sodium dithionite.

Chlorine and hypochlorite

Chlorine replaces hydrogen on the aromatic rings of lignin via aromatic substitution, oxidizes pendant groups to carboxylic acids and adds across carbon carbon double bonds in the lignin sidechains. Chlorine also attacks cellulose, but this reaction occurs predominantely at pH 7, where un-ionized hypochlorous acid, HClO, is the main chlorine species in solution. To avoid excessive cellulose degradation, chlorination is carried out at pH <1.5.

$$Cl_2 + H_2O \rightleftharpoons H^+ + Cl^" + HClO$$

At pH >8 the dominant species is hypochlorite, ClO^{-}, which is also useful for lignin removal. Sodium hypochlorite can be purchased or generated *in situ* by reacting chlorine withsodium hydroxide.

$$2\,NaOH + Cl_2 \rightleftharpoons NaOCl + NaCl + H_2O$$

The main objection to the use of chlorine for bleaching pulp is the large amounts of soluble organochlorine compounds produced and released into the environment.

Chlorine dioxide

Chlorine dioxide, ClO_2 is an unstable gas with moderate solubility in water. It is usually generated in an aqueous solution and used immediately because it decomposes and is explosive in higher concentrations. It is produced by reacting sodium chlorate with a reducing agent like sulfur dioxide.

$$2\,NaClO_3 + H_2SO_4 + SO_2 \rightarrow 2\,ClO_2 + 2\,NaHSO_4$$

Chlorine dioxide is sometimes used in combination with chlorine, but it is used alone in ECF (elemental chlorine-free) bleaching sequences. It is used at moderately acidic pH (3.5 to 6). The use of chlorine dioxide minimizes the amount of organochlorine compounds produced. Chlorine dioxide (ECF technology) currently is the most important bleaching method worldwide. About 95% of all bleached Kraft pulp is made using chlorine dioxide in ECF bleaching sequences.

Extraction or washing

All bleaching agents used to delignify chemical pulp, with the exception of sodium dithionite, break lignin down into smaller, oxygen-containing molecules. These breakdown products are generally soluble in water, especially if the pH is greater than 7 (many of the products are carboxylic acids). These materials must be removed between bleaching stages to avoid excessive use of bleaching chemicals since many of these smaller molecules are still susceptible to oxidation. The need to minimize water use in modern pulp mills has driven the development of equipment and techniques for the efficient use of available water.

Oxygen

Oxygen exists as a ground state triplet state, which is relatively unreactive and needs free radicals or very electron-rich substrates such as deprotonated lignin phenolic groups.

The production of these phenoxide groups requires that delignification with oxygen be carried out under very basic conditions (pH >12). The reactions involved are primarily single electron (radical) reactions. Oxygen opens rings and cleaves sidechains giving a complex mixture of small oxygenated molecules. Transition metal compounds, particularly those of iron, manganese and copper, which have multiple oxidation states, facilitate many radical reactions and impact oxygen delignification. While the radical reactions are largely responsible for delignification, they are detrimental to cellulose. Oxygen-based radicals, especially hydroxyl radicals, HO•, can oxidize hydroxyl groups in the cellulose chains to ketones, and under the strongly basic conditions used in oxygen delignification, these compounds undergo reverse aldol reactions leading to cleavage of cellulose chains. Magnesium salts are added to oxygen delignification to help preserve the cellulose chains, but mechanism of this protection has not been confirmed.

Hydrogen peroxide

Using hydrogen peroxide to delignify chemical pulp requires more vigorous conditions than for brightening mechanical pulp. Both pH and temperature are higher when treating chemical pulp. The chemistry is very similar to that involved in oxygen delignification, in terms of the radical species involved and the products produced. Hydrogen peroxide is sometimes used with oxygen in the same bleaching stage and this is give the letter designation Op in bleaching sequences. Metal ions, particularly manganese catalyze the decomposition of hydrogen peroxide, so some improvement in the efficiency of peroxide bleaching can be achieved if the metal levels are controlled.

Ozone

Ozone is a very powerful oxidizing agent and the biggest challenge in using it to bleach wood pulp is to get sufficient selectivity so that the desirable cellulose is not degraded. Ozone reacts with the carbon carbon double bonds in lignin, including those within aromatic rings. In the 1990s ozone was touted as good reagent to allow pulp to be bleached without any chlorine-containing chemicals (totally chlorine-free, TCF). The emphasis has changed and ozone is seen as an adjunct to chlorine dioxide in bleaching sequences not using any elemental chlorine (elemental chlorine-free, ECF). Over twenty-five pulp mills worldwide have installed equipment to generate and use ozone.

Chelant wash

The effect of transition metals on some of the bleaching stages has already been mentioned. Sometimes it is beneficial to remove some of these metal ions from the pulp by washing the pulp with a chelating agent such as EDTA or DTPA. This is more common in TCF bleaching sequences for two reasons: the acidic chlorine or chlorine dioxide stages tend to remove metal ions (metal ions usually being more soluble at lower pH) and TCF stages rely more heavily on oxygen-based bleaching agents, which are more susceptible to the detrimental effects of these metal ions. Chelant washes are usually carried out at or near pH 7. Lower pH solutions are more effective at removing transition metals, but also remove more of the beneficial metal ions, especially magnesium

Other bleaching agents

A variety of more exotic bleaching agents have been used on chemical pulps. They include peroxyacetic acid, peroxyformic acid, potassium peroxymonosulfate(Oxone), dimethyldioxirane, which is generated *in situ* from acetone and potassium peroxymonosulfate, and peroxymonophosphoric acid

Enzymes like xylanase have been used in pulp bleaching to increase the efficiency of other bleaching chemicals. It is believed that xylanase does this by cleaving lignin-xylan bonds to make lignin more accessible to other reagents. It is possible that other enzymes such as thôse found in fungi that degrade lignin may be useful in pulp bleaching.

ENVIRONMENTAL CONSIDERATIONS

Bleaching mechanical pulp is not a major cause for environmental concern since most of the organic material is retained in the pulp, and the chemicals used (hydrogen peroxideand sodium dithionite) produce benign byproducts (water and sodium sulfate (finally), respectively).

However, the bleaching of chemical pulps has the potential to cause significant environmental damage, primarily through the release of organic materials into waterways. Pulp mills are almost always located near large bodies of water because they require substantial quantities of water for their processes. An increased public awareness of environmental issues from the 1970s and 1980s, as evidenced by the formation of organizations like Greenpeace, influenced the pulping industry and governments to address the release of these materials into the environment.

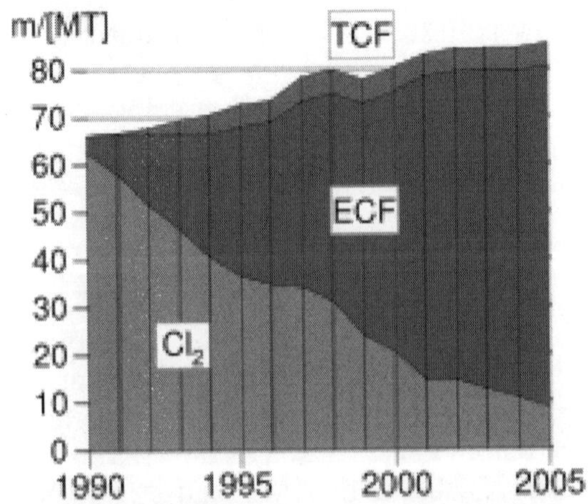

Worldwide pulp production by type of bleaching used: Chlorine (Cl_2), Elementary Chlorine Free (ECF) and Total Chlorine Free (TCF).

Conventional bleaching using elemental chlorine produces and releases into the environment large amounts of chlorinated organic compounds, including chlorinated dioxins. Dioxins are recognized as a persistent environmental pollutant, regulated internationally by the Stockholm Convention on Persistent Organic Pollutants.

Dioxins are highly toxic, and health effects on humans include reproductive, developmental, immune and hormonal problems. They are known to be carcinogenic. Over 90% of human exposure is through food, primarily meat, dairy, fish and shellfish, as dioxins accumulate in the food chain in the fatty tissue of animals.

As a result, from the 1990s onwards, the use of elemental chlorine in the delignification process was substantially reduced and replaced with ECF (Elemental Chlorine Free) and TCF (Totally Chlorine Free) bleaching processes. In 2005, elemental chlorine was used in 19-20% of kraft pulp production globally, down from over 90% in 1990. 75% of kraft pulp used ECF, with the remaining 5-6% using TCF.Most TCF pulp is produced in Sweden and Finland for sale in Germany, all markets with a high level of environmental awareness. In 1999, TCF pulp represented 25% of the European market.

TCF bleaching, by removing chlorine from the process, reduces chlorinated organic compounds to background levels in pulp mill effluent. ECF bleaching can substantially reduce but not fully eliminate chlorinated organic compounds, including dioxins, from effluent. While modern ECF plants can achieve chlorinated organic compounds (AOX) emissions of less than 0.05 kg per tonne of pulp produced, most do not achieve this level of emissions. Within the EU, the average chlorinated organic compound emissions for ECF plants is 0.15 kg per tonne.

However, there has been disagreement about the comparative environmental effects of ECF and TCF bleaching. Some researchers found that there is no environmental difference between ECF and TCF while others concluded that among ECF and TCF effluents before and after secondary treatment, TCF effluents are the least toxic.

THE MILL

Much of the information about the technology in following subsections is from the book by C.J. Biermann. The chemistry of the various pulping processes can be found in Sjöström's book.

Preparation of fibre source

The most common fibre source for pulp mills is pulpwood. Other common sources are bagasse and fibre crops. The first step in all mills using wood (trees) as the fibre source is to remove the bark. Bark contains relatively few usable fibres and darkens the pulp. The removed bark is burned, along with other unusable plant material, to generate steam to run the mill. Almost all wood is then chippedbefore being processed further in order to free the fibres.

Removal of the bark is done in a *barker* (or *debarker*). The bark adhesion is about 3–5 kg/cm^2 in the growing season (summer) and 2-3 times higher in the dormant season (winter). The bark of frozen logs is even more difficult to remove.

In chemical pulp mills, the bark introduces unwanted contaminants such as calcium, silica and aluminium that cause scaling and give an extra loading for the chemical recovery system. Birch bark contains betulin, a terpenoid that easily creates deposits in a pulp mill.

Mechanical pulp mills

The earliest mills used sandstone grinding rollers to break up small wood logs called "bolts", but the use of natural stone ended in the 1940s with the introduction of manufactured stones with embedded silicon carbide or aluminum oxide. The pulp made by this process is known as "stone groundwood" pulp (SGW). If the wood is ground in a pressurized,sealed grinder the pulp is classified as "pressure groundwood" (PGW) pulp. Most modern mills use chips rather than logs and ridged metal discs called refiner plates instead of grindstones. If the chips are just ground up with the plates, the pulp is called "refiner mechanical" pulp (RMP), if the chips are steamed while being refined the pulp is called "thermomechanical" pulp (TMP). Steam treatment significantly reduces the total energy needed to make the pulp and decreases the damage (cutting) to fibers. Mechanical pulp mills use large amounts of energy, mostly electricity to power motors which turn the grinders. A rough estimate of the electrical energy needed is 10,000 megajoules (MJ) per tonne of pulp (2,750 kWh per tonne)

Chemical pulp mills

Chemical pulping processes such as the kraft (or sulphate) process and the sulphite process remove much of the hemicelluloses and lignin. The kraft process does less damage to the cellulose fibres than the sulphite process, thereby producing stronger fibres, but the sulphite process makes pulp that is easier to bleach. The chemical pulping processes use a combination of high temperature and alkaline (kraft) or acidic (sulphite) chemicals to break the chemical bonds of the lignin.

The material fed into the digester must be small enough to allow the pulping liquor to penetrate the pieces completely. In the case of wood, the logs are chipped and the chips screened so that what is fed to the digester is a uniform size. The oversize chips are rechipped or used as fuel, sawdust is burned. The screened chips or cut plant material (bamboo, kenaf, etc.) goes to the digester where it is mixed an aqueous solution of the pulping chemicals, then heated with steam. In the kraft process the pulping chemicals are sodium hydroxideand sodium sulphide and the solution is known as white liquor. In the sulphite process the pulping chemical is a mixture of metal (sodium, magnesium, potassium or calcium) or ammonium sulphite or sulphite.

After several hours in the digester, the chips or cut plant material breaks down into a thick porridge-like consistency and is "blown" or squeezed from the outlet of the digester through an airlock. The sudden change in pressure results in a rapid expansion of the fibres, separating the fibres even more. The resulting fibre suspension in water solution is called "brown stock".

Brown stock washers, using countercurrent flow, remove the spent cooking chemicals and degraded lignin and hemicellulose. The extracted liquid, known as black liquor in the kraft process, and red or brown liquor in the sulphite processes, is concentrated, burned and the sodium and sulphur compounds recycled in the recovery process.Lignosulphonates are a useful byproduct recovered from the spent liquor in the sulphite process. The clean pulp (stock) can be bleached in the bleach plant or left unbleached, depending on the end use. The stock is sprayed onto the pulp machine wire, water drains off, more water is removed by pressing the sheet of fibres, and the sheet is then dried. At this point the sheets of pulp are several millimetres thick and have a coarse surface: it is not yet paper. The dried pulp is cut, stacked, bailed and shipped to another facility for whatever further process is needed.

Bleached kraft pulp and bleached sulphite pulp are used to make high quality, white printing paper. One of the most visible uses for unbleached kraft pulp is to make brown paper shopping bags and wrapping paper where strength is particularly important. A special grade of bleached sulphite pulp, known as dissolving pulp, is used to make cellulose derivatives such as methylcellulose which are used in a wide range of everyday products from laxatives to baked goods to wallpaper paste.

Chemi-mechanical pulp mills

Some mills pretreat wood chips or other plant material like straw with sodium carbonate, sodium hydroxide, sodium sulfite, and other chemical prior to refining with equipment similar to a mechanical mill. The conditions of the chemical treatment are much less vigorous (lower temperature, shorter time, less extreme pH) than in a chemical pulping process, since the goal is to make the fibers easier to refine, not to remove lignin as in a fully chemical process. Pulps made using these hybrid processes are known as chemi-thermomechanical pulps (CTMP). Sometimes a CTMP mill is located on the same site as a kraft mill so that the effluent from the CTMP mill can be treated in the kraft recovery process to regenerate the inorganic pulping chemicals.

SCHEDULING IN PULP MILLS

The pulp process involves many production stages, usually coupled with intermediate storage tanks. As each stage has a different reliability and bottlenecks may vary from day to day, scheduling a pulp mill needs to take into account these bottlenecks and the probability of a disturbance or breakdown. Each stage also may have different decision variables, such as steam / water / chemical input, etc. Finally, scheduling needs to consider fuel optimisation and CO_2 emissions, because part of the energy requirements may be met from fossil-fuel boilers. The overall aim is to maximise production at minimum cost.

PULP AND PAPER INDUSTRY

EVOLUTION AND STRUCTURE OF THE INDUSTRY

Papermaking is thought to have originated in China in about 100 A.D. using rags, hemp and grasses as the raw material, and beating against stone mortars as the original fibre separation process. Although mechanization increased over the intervening years, batch production methods and agricultural fibre sources remained in use until the 1800s. Continuous papermaking machines were patented at the turn of that century. Methods for pulping wood, a more abundant fibre source than rags and grasses, were developed between 1844 and 1884, and included mechanical abrasion as well as the soda, sulphite, and sulphate (kraft) chemical methods. These changes initiated the modern pulp and paper manufacturing era.

There has been a trend in recent years for pulp and paper operations to become part of large, integrated forest product companies. These companies have control of forest harvesting operations, lumber milling, pulp and paper manufacturing, as well as converting operations. This structure ensures that the company has an ongoing source of fibre, efficient use of wood waste and assured buyers, which often leads to increased market share. Integration has been operating in tandem with increasing concentration of the industry into fewer companies and increasing globalization as companies pursue international investments.

The financial burden of plant development in this industry has encouraged these trends to allow economies of scale. Some companies have now reached production levels of 10 million tonnes, similar to the output of countries with the highest production. Many companies are multinational, some with plants in 20 or more countries worldwide. However, even though many of the smaller mills and companies are disappearing, the industry still has hundreds of participants. As an illustration, the top 150 companies account for two-thirds of pulp and paper output and only one-third of the industry's employees.

ECONOMIC IMPORTANCE

The manufacture of pulp, paper and paper products ranks among the world's largest industries. Mills are found in more than 100 countries in every region of the world, and directly employ more than 3.5 million people. The major pulp and paper producing nations include the United States, Canada, Japan, China, Finland, Sweden, Germany, Brazil and France.

Table : Employment and production in pulp, paper, and paperboard operations in 1994, selected countries.

Country *	Number employed in industry	Pulp		Paper and paperboard	
		Number of mills	Production (1,000 tonnes)	Number of mills	Production (1,000 tonnes)
Austria	10,000	11	1,595	28	3,603
Bangladesh	15,000	7	84	17	160
Brazil	70,000	35	6,106	182	5,698
Canada	64,000	39	24,547	117	18,316
China	1,500,000	8,000	17,054	10,000	21,354
Czech Republic	18,000	9	516	32	662
Finland	37,000	43	9,962	44	10,910
Former USSR**	178,000	50	3,313	161	4,826
France	48,000	20	2,787	146	8,678
Germany	48,000	19	1,934	222	14,458
India	300,000	245	1,400	380	2,300
Italy	26,000	19	535	295	6,689
Japan	55,000	49	10,579	442	28,527
Korea, Republic of	60,000	5	531	136	6,345
Mexico	26,000	10	276	59	2,860
Pakistan	65,000	2	138	68	235
Poland**	46,000	5	893	27	1,343
Romania	25,000	17	202	15	288
Slovakia	14,000	3	304	6	422
South Africa	19,000	9	2,165	20	1,684
Spain	20,180	21	626	141	5,528
Sweden	32,000	49	10,867	50	9,354
Taiwan	18,000	2	326	156	4,199
Thailand	12,000	3	240	45	1,664
Turkey	12,000	11	416	34	1,102
United Kingdom	25,000	5	626	99	5,528
United States	230,000	190	58,724	534	80,656
Total worldwide	approx 3,500,000	9,100	171,479	14,260	268,551

* Countries included if more than 10,000 people were employed in the industry.
** Data for 1989/90 (ILO 1992).

Every country is a consumer. Worldwide production of pulp, paper and paperboard was about 400 million tonnes in 1993. Despite predictions of decreased paper use in the face of the electronic age, there has been a fairly steady 2.5% annual rate of growth in production since 1980. In addition to its economic benefits, the consumption of paper has cultural value resulting from its function in the recording and dissemination of information. Because of this, pulp and paper consumption rates have been used as an indicator of a nation's socioeconomic development.

The main source of fibre for pulp production over the last century has been wood from temperate coniferous forests, though more recently the use of tropical and boreal woods has been increasing. Because forested regions of the world are generally sparsely populated, there tends to be a dichotomy between the producing and using areas of the world. Pressure from environmental groups to preserve forest resources by using recycled paper stocks, agricultural crops and short-rotation plantation forests as fibre sources may change the distribution of pulp and paper production facilities throughout the world over the coming decades. Other forces, including increased paper consumption in the developing world and globalization, are also expected to play a role in relocating the industry.

CHARACTERISTICS OF THE WORKFORCE

The size of the workforce directly employed in pulp and paper production and converting operations in 27 countries, which together represent about 85% of world pulp and paper employment and over 90% of mills and production. In countries which consume most of what they produce (e.g., United States, Germany, France), converting operations provide two jobs for every one in pulp and paper production.

The labour force in the pulp and paper industry mainly holds full-time jobs within traditional management structures, though some mills in Finland, the United States and elsewhere have had success with flexible working hours and self-managed job-rotation teams. Because of their high capital costs, most pulping operations run continuously and require shift work; this is not true of converting plants. Working hours vary with the patterns of employment prevalent in each country, with a range from about 1,500 to more than 2,000 hours per year. In 1991, incomes in the industry ranged from US$1,300 (unskilled workers in Kenya) to US$70,000 per year (skilled production personnel in the United States) (ILO 1992). Male workers predominate in this industry, with women usually representing only 10 to 20% of the labour force. China and India may form the upper and lower ends of the range with 35% and 5% women respectively.

Management and engineering personnel at pulp and paper mills usually have university-level training. In European countries, most of the skilled blue-collar workforce (e.g., papermakers) and many of the unskilled workforce have had several years of trade-school education. In Japan, formal in-house training and upgrading is the norm; this approach is being adopted by some Latin American and North American companies. However, in many operations in North America and in the developing world, informal on-the-job training is more common for blue-collar jobs. Surveys have shown that, in some operations, many workers have literacy problems and are poorly prepared for the life-long learning required in the dynamic and potentially hazardous environment of this industry.

The capital costs of building modern pulp and paper plants are extremely high (e.g., a bleached kraft mill employing 750 people might cost US$1.5 billion to build; a chemi-thermomechanical pulp (CTMP) mill employing 100 people might cost US$400 million), so there are great economies of scale with high-capacity facilities. New and refitted plants usually use mechanized and continuous processes, as well as electronic monitors and computer controls. They require relatively few employees per unit production (e.g., 1 to 1.2 working hours per tonne of pulp in new Indonesian, Finnish and Chilean mills). Over the last 10 to 20 years, output per employee has increased as a result of incremental advances in technology. The newer equipment allows easier change-overs between product runs, lower inventories and customer-driven just-in-time production. Productivity gains have resulted in job losses in many producing nations in the developed world. However, there have been increases in employment in developing countries, where new mills being constructed, even if sparsely staffed, represent new forays into the industry.

From the 1970s to 1990, there was a decline of about 10% in the proportion of blue-collar jobs in European and North American operations, so that they now represent between 70 and 80% of the workforce (ILO 1992). The use of contract labour for mill construction, maintenance and wood-harvesting operations has been increasing; many operations have reported that 10 to 15% of their on-site workforce are contractors.

FIBRE SOURCES FOR PULP AND PAPER

The basic structure of pulp and paper sheets is a felted mat of cellulose fibres held together by hydrogen bonds. Cellulose is a polysaccharide with 600 to 1,500 repeated sugar units. The fibres have high tensile strength, will absorb the additives used to modify pulp into paper and board products, and are supple, chemically stable and white. The purpose of pulping is to separate cellulose fibres from the other components of the fibre source. In the case of wood, these include hemicelluloses (with 15 to 90 repeated sugar units), lignins (highly polymerized and complex, mainly phenyl propane units; they act as the "glue" that cements the fibres together), extractives (fats, waxes, alcohols, phenols, aromatic acids, essential oils, oleoresins, stearols, alkaloids and pigments), and minerals and other inorganics. The relative proportions of these components vary according to the fibre source.

Table : Chemical constituents of pulp and paper fibre sources (%)

	Softwoods	Hardwoods	Straw	Bamboo	Cotton
Carbohydrates					
α-cellulose	38–46	38–49	28–42	26–43	80–85
Hemicellulose	23–31	20–40	23–38	15–26	nd
Lignin	22–34	16–30	12–21	20–32	nd
Extractives	1–5	2–8	1–2	0.2–5	nd
Minerals and other inorganics	0.1–7	0.1–11	3–20	1–10	0.8–2

nd = no data available.

Coniferous and deciduous trees are the major fibre sources for pulp and paper. Secondary sources include straws from wheat, rye and rice; canes, such as bagasse; woody stalks from bamboo, flax and hemp; and seed, leaf or bast fibres, such as cotton, abaca and sisal. The majority of pulp is made from virgin fibre, but recycled paper accounts for an increasing proportion of production, up from 20% in 1970 to 33% in 1991.

WOOD HANDLING

Wood may arrive at a pulp mill woodyard in the form of raw logs or as chips from a lumber mill. Some pulp mill operations have on-site sawmills (often called "woodrooms") which produce both marketable lumber and stock for the pulp mill.

The wood preparation area of a pulp mill has several basic functions: to receive and meter the wood supply to the pulping process at the rate demanded by the mill; to prepare the wood so that it meets the mill's feed specifications for species, cleanliness and dimensions; and to collect any material rejected by the previous operations and send it to final disposal. Wood is converted into chips or logs suitable for pulping in a series of steps which may include debarking, sawing, chipping and screening.

Logs are debarked because bark contains little fibre, has a high extractives content, is dark, and often carries large quantities of grit. Debarking can be done hydraulically with high-pressure water jets, or mechanically by rubbing logs against each other or with metal cutting tools. Hydraulic debarkers may be used in coastal areas; however, the effluent generated is difficult to treat and contributes to water pollution.

Debarked logs may be sawn into short lengths (1 to 6 metres) for stone groundwood pulping or chipped for refiner mechanical or chemical pulping methods. Chippers tend to produce chips with a considerable size range, but pulping requires chips of very specific dimensions to ensure constant flow through refiners and uniform cooking in digesters. Chips are therefore passed over a series of screens whose function is to separate chips on the basis of length or thickness. Oversized chips are rechipped, while undersized chips are either used as waste fuel or are metered back into the chip flow.

The requirements of the particular pulping process and chip conditions will dictate the duration of chip storage. Depending on fibre supply and mill demand, a mill will maintain a 2 to 6 week unscreened chip inventory, usually in large outdoor chip piles. Chips may degrade through auto-oxidation and hydrolysis reactions or fungal attack of the wood components. In order to avoid contamination, short-term inventories (hours to days) of screened chips are stored in chip silos or bins. Chips for sulphite pulping may be stored outside for several months to allow volatilization of extractives which may cause problems in subsequent operations. Chips used in kraft mills where turpentine and tall oil are recovered as commercial products typically proceed directly to pulping.

Fig. Chip storage area with front end loaders

PULPING

Pulping is the process by which the bonds within the wood structure are ruptured either mechanically or chemically. Chemical pulps can be produced by either alkaline (i.e., sulphate or kraft) or acidic (i.e., sulphite) processes. The highest proportion of pulp is produced by the sulphate method, followed by mechanical (including semi-chemical, thermomechanical and mechanical) and sulphite methods. Pulping processes differ in the yield and quality of the product, and for chemical methods, in the chemicals used and the proportion that can be recovered for reuse.

Mechanical Pulping

Mechanical pulps are produced by grinding wood against a stone or between metal plates, thereby separating the wood into individual fibres. The shearing action breaks cellulose fibres, so that the resulting pulp is weaker than chemically separated pulps. The lignin connecting cellulose to hemicellulose is not dissolved; it merely softens, allowing the fibres to be ground out of the wood matrix. The yield (proportion of original wood in pulp) is usually greater than 85%. Some mechanical pulping methods also use chemicals (i.e., the chemi-mechanical pulps); their yields are lower since they remove more of the non-cellulosic materials.

In stone groundwood pulping (SGW), the oldest and historically most common mechanical method, fibres are removed from short logs by pressing them against a rotating abrasive cylinder. In refiner mechanical pulping, which gained popularity after it became commercially viable in the 1960s, wood chips or sawdust are fed through the centre of a disc refiner, where they are shredded into finer pieces as they are pushed out through progressively narrower bars and grooves. A modification of RMP is thermomechanical pulping (TMP), in which the chips are steamed before and during refining, usually under pressure.

Figure : Refiner mechanical pulping

f*****

One of the earliest methods of producing chemi-mechanical pulps involved pre-steaming logs before boiling them in chemical pulping liquors, then grinding them in stone grinders to produce "chemi-groundwood" pulps. Modern chemi-mechanical pulping uses disc refiners with chemical treatment (e.g., sodium bisulphite, sodium hydroxide) either prior to, during or after refining. Pulps produced in this manner are referred to either as chemi-mechanical pulps (CMP) or chemi-thermomechanical pulps (CTMP), depending on whether refining was carried out at atmospheric or elevated pressure. Specialized variations of CTMP have been developed and patented by a number of organizations.

CHEMICAL PULPING AND RECOVERY

Chemical pulps are produced by chemically dissolving the lignin between the wood fibres, thereby enabling the fibres to separate relatively undamaged. Because most of the non-fibrous wood components are removed in these processes, yields are usually in the order of 40 to 55%.

In chemical pulping, chips and chemicals in aqueous solution are cooked together in a pressure vessel which can be operated on a batch or continuous basis. In batch cooking, the digester is filled with chips through a top opening, the digestion chemicals are added, and the contents cooked at elevated temperature and pressure. Once the cook is complete, the pressure is released, "blowing" the delignified pulp out of the digester and into a holding tank. The sequence is then repeated. In continuous digesting, pre-steamed chips are fed into the digester at a continuous rate. Chips and chemicals are mixed together in the impregnation zone at the top of the digester and then proceed through the upper cooking zone, the lower cooking zone, and the washing zone before being blown into the blow tank.

Fig. Continuous kraft digestor, with chip conveyor under construction

The digesting chemicals are recovered in most chemical pulping operations today. The principal objectives are to recover and reconstitute digestion chemicals from the spent cooking liquor, and to recover heat energy by burning the dissolved organic material from the wood. The resulting steam and electricity supplies some, if not all, of the mill's energy needs.

SULPHATE PULPING AND RECOVERY

The sulphate process produces a stronger, darker pulp than other methods and requires chemical recovery to compete economically. The method evolved from soda pulping (which uses only sodium hydroxide for digestion) and began to gain prominence in the industry from the 1930s to 1950s with the development of chlorine dioxide bleaching and chemical recovery processes, which also produced steam and power for the mill. The development of corrosion-proof metals, such as stainless steel, to handle the acidic and alkaline pulp mill environments also played a role.

The cooking mixture (white liquor) is sodium hydroxide (NaOH, "caustic") and sodium sulphide (Na_2S). Modern kraft pulping is usually carried out in continuous digesters often lined with stainless steel. The temperature of the digester is raised slowly to approximately 170°C and held at that level for approximately 3 to 4 hours.

The pulp (called brown stock because of its colour) is screened to remove uncooked wood, washed to remove the spent cooking mixture (now black liquor), and sent either to the bleach plant or to the pulp machine room. Uncooked wood is either returned to the digester or sent to the power boiler to be burned.

The black liquor collected from the digester and brown stock washers contains dissolved organic material whose exact chemical composition depends on the wood species pulped and the cooking conditions. The liquor is concentrated in evaporators until it contains less than 40% water, then sprayed into the recovery boiler. The organic component is consumed as fuel, generating heat which is recovered in the upper section of the furnace as high-temperature steam.

The unburned inorganic component collects at the bottom of the boiler as a molten smelt. The smelt flows out of the furnace and is dissolved in a weak caustic solution, producing "green liquor" containing primarily dissolved Na_2S and sodium carbonate (Na_2CO_3). This liquor is pumped to a recausticizing plant, where it is clarified, then reacted with slaked lime ($Ca(OH)_2$), forming NaOH and calcium carbonate ($CaCO_3$). The white liquor is filtered and stored for subsequent use. $CaCO_3$ is sent to a lime kiln, where it is heated to regenerate lime (CaO).

SULPHITE PULPING AND RECOVERY

Sulphite pulping dominated the industry from the late 1800s to the mid-1900s, but the method used during this era was limited by the types of wood which could be pulped and the pollution created by discharging untreated waste cooking liquor into waterways. Newer methods have overcome many of these problems, but sulphite pulping is now a small segment of the pulp market. Although sulphite pulping usually uses acid digestion, both neutral and basic variations exist.

The cooking liquor of sulphurous acid (H_2SO_3) and bisulphite ion (HSO_3–) is prepared on-site. Elemental sulphur is burned to produce sulphur dioxide (SO_2), which is passed up through an absorption tower that contains water and one of four alkaline bases ($CaCO_3$, the original sulphite base, Na_2CO_3, magnesium hydroxide ($Mg(OH)_2$) or ammonium hydroxide (NH_4OH)) which produce the acid and ion and control their proportions. Sulphite pulping is usually carried out in brick-lined batch digesters. To avoid unwanted reactions, the digester is heated slowly to a maximum temperature of 130 to 140°C and the chips are cooked for a long time (6 to 8 hours). As the digester pressure increases, gaseous sulphur dioxide (SO_2) is bled off and remixed with the raw cooking acid. When approximately 1 to 1.5 hours of cooking time remains, heating is discontinued and the pressure is decreased by bleeding off gas and steam. The pulp is blown into a holding tank, then washed and screened.

The spent digestion mixture, called red liquor, can be used for heat and chemical recovery for all but calcium-bisulphite-base operations. For ammonia-base sulphite pulping, the dilute red liquor is first stripped to remove residual SO_2, then concentrated and burned.

The flue gas containing SO_2 is cooled and passed through an absorption tower where fresh ammonia combines with it to regenerate the cooking liquor. Finally, the liquor is filtered, fortified with fresh SO_2 and stored. The ammonia cannot be recovered because it is converted into nitrogen and water in the recovery boiler.

In magnesium-base sulphite pulping, burning the concentrated pulping liquor gives magnesium oxide (MgO) and SO_2, which are easily recovered. No smelt is produced in this process; rather MgO is collected from the flue gas and slaked with water to produce magnesium hydroxide ($Mg(OH)_2$). SO_2 is cooled and combined with the $Mg(OH)_2$ in an absorption tower to reconstitute the cooking liquor. The magnesium bisulphite ($Mg(HSO_3)_2$) is then fortified with fresh SO_2 and stored. Recovery of 80 to 90% of the cooking chemicals is possible.

Recovery of sodium-base sulphite cooking liquor is more complicated. Concentrated spent liquor is incinerated, and approximately 50% of the sulphur is converted into SO_2. The remainder of the sodium and sulphur is collected at the bottom of the recovery boiler as a smelt of Na_2S and Na_2CO_3. The smelt is dissolved to produce green liquor, which is converted to sodium bisulphite ($NaHSO_3$) in several steps. The $NaHSO_3$ is fortified and stored. The regeneration process produces reduced sulphur gases, in particular hydrogen sulphide (H_2S).

BLEACHING

Bleaching is a multi-stage process that refines and brightens raw pulp. The objective is to dissolve (chemical pulps) or modify (mechanical pulps) the brown-coloured lignin that was not removed during pulping, while maintaining the integrity of the pulp fibres. A mill produces customized pulp by varying the order, concentration and reaction time of the bleaching agents.

Each bleaching stage is defined by its bleaching agent, pH (acidity), temperature and duration.

After each bleaching stage, the pulp may be washed with caustic to remove spent bleaching chemicals and dissolved lignin before it progresses to the next stage. After the last stage, the pulp is pumped through a series of screens and cleaners to remove any contaminants such as dirt or plastic. It is then concentrated and conveyed to storage.

Table : Bleaching agents and their conditions of use

	Symbol	Concentration of agent (%)	pH	Consistency* (%)	Temperature (°C)	Time (h)
Chlorine (Cl_2)	C	2.5–8	2	3	20–60	0.5–1.5
Sodium hydroxide (NaOH)	E	1.5–4.2	11	10–12	<80	1–2
Chlorine dioxide (ClO_2)	D	approx 1	0–6	10–12	60–75	2–5
Sodium hypochlorite (NaOCl)	H	1–2	9–11	10–12	30–50	0.5–3
Oxygen (O_2)	O	1.2–1.9	7–8	25–33	90–130	0.3–1
Hydrogen peroxide (H_2O_2)	P	0.25	10	12	35–80	4
Ozone (O_3)	Z	0.5–3.5	2–3	35–55	20–40	<0.1
Acid washing (SO_2)	A	4–6	1.8–5	1.5	30–50	0.25
Sodium dithionite (NaS_2O_4)	Y	1–2	5.5–8	4–8	60–65	1–2

* Concentration of fibre in water solution.

Historically, the most common bleaching sequence used to produce market-grade bleached kraft pulp is based on the five-stage CEDED process. The first two stages of bleaching complete the delignification process and are considered extensions of pulping. Because of environmental concerns about chlorinated organics in pulp mill effluents, many mills substitute chlorine dioxide (ClO_2) for a portion of the chlorine (Cl_2) used in the first bleaching stage (C_DEDED) and use oxygen (O_2) pre-treatment during the first caustic extraction (C_DE_ODED). The current trend in Europe and North America is towards complete substitution with ClO_2 (e.g., DEDED) or elimination of both Cl_2 and ClO_2. Where ClO_2 is used, sulphur dioxide (SO_2) is added during the final washing stage as an "antichlor" to stop the ClO_2 reaction and to control the pH. Newly developed chlorine-free bleaching sequences (e.g., OAZQP, OQPZP, where Q = chelation) use enzymes, O_2, ozone (O_3), hydrogen peroxide (H_2O_2), peracids and chelating agents such as ethylene diamine tetracetic acid (EDTA). Totally chlorine-free bleaching had been adopted at eight mills worldwide by 1993. Because these newer methods eliminate the acidic bleaching steps, acid washing is a necessary addition to the initial stages of kraft bleaching to allow removal of metals bound to the cellulose.

Sulphite pulps are generally easier to bleach than kraft pulps because of their lower lignin content. Short bleaching sequences (e.g., CEH, DCEHD, P, HP, EPOP) can be used for most paper grades. For dissolving-grade sulphite pulps used in the production of rayon, cellophane and so on, both hemicellulose and lignin are removed, requiring more complex bleaching sequences (e.g., C_1C_2ECHDA). The final acid wash is both for metal control and antichlor purposes. The effluent load for dissolving-grade sulphite pulps is much greater because so much of the raw wood is consumed (typical yield 50%) and more water is used.

The term brightening is used to describe bleaching of mechanical and other high-yield pulps, because they are whitened by destroying chromophoric groups without dissolving the lignin. Brightening agents include H_2O_2 and/or sodium hydrosulphite (NaS_2O_4). Historically, zinc hydrosulphite (ZnS_2O_4) was commonly used, but has been largely eliminated because of its toxicity in effluent. Chelating agents are added before bleaching to neutralize any metal ions, thereby preventing the formation of coloured salts or the decomposition of H_2O_2. The effectiveness of mechanical pulp bleaching depends on the species of wood. Hardwoods (e.g., poplar and cottonwood) and softwoods (e.g., spruce and balsam) that are low in lignin and extractives can be bleached to a higher brightness level than the more resinous pine and cedar.

RECYCLED PAPER OPERATIONS

The use of waste or recycled paper as the raw material for pulp production has increased during the last several decades, and some paper plants depend almost completely on waste paper. In some countries, waste paper is separated from other household waste at the source before it is collected. In other countries separation by grade (e.g., corrugated board, newsprint, high-grade paper, mixed) takes place in special recycling plants.

Recycled paper can be repulped in a relatively mild process which uses water and sometimes NaOH. Small metal pieces and plastics may be separated during and/or after repulping, using a debris rope, cyclones or centrifugation. Filling agents, glues and resins are removed in a cleaning stage by blowing air through the pulp slurry, sometimes with the addition of flocculating agents. The foam contains the unwanted chemicals and is removed. The pulp can be de-inked using a series of washing steps which may or may not include the use of chemicals (i.e., surfactant fatty acid derivatives) to dissolve remaining impurities, and bleaching agents to whiten the pulp. Bleaching has the disadvantage that it may reduce fibre length and therefore lessen final paper quality. The bleaching chemicals used in recycled pulp production are usually similar to those used in brightening operations for mechanical pulps. After the repulping and de-inking operations, sheet production follows in a manner very similar to that using virgin fibre pulp.

SHEET PRODUCTION AND CONVERTING: MARKET PULP, PAPER, PAPERBOARD

End products of pulp and paper mills depend on the pulping process, and may include market pulp and various types of paper or paperboard products. For example, the relatively weak mechanical pulp is converted into single-use products such as newspapers and tissue. Kraft pulp is converted into multi-use paper products such as high-quality writing paper, books and grocery bags. Sulphite pulp, which is primarily cellulose, can be used in a series of diverse end-products including specialty paper, rayon, photographic film, TNT, plastics, adhesives, and even ice cream and cake mixes. Chemi-mechanical pulps are exceptionally stiff, ideal for the structural support needed for corrugated container board. The fibres in pulp from recycled paper are usually shorter, less flexible and less water permeable, and can therefore not be used for high-quality paper products. Recycled paper is therefore mainly used for the production of soft paper products like tissue paper, toilet paper, paper towelling and napkins.

To produce market pulp, the pulp slurry is usually screened once more and its consistency adjusted (4 to 10%) before it is ready for the pulp machine. The pulp is then spread onto a travelling metal screen or plastic mesh (known as the "wire") at the "wet end" of the pulp machine, where the operator monitors the speed of the moving wire and the water content of the pulp. Water and filtrate are drawn through the wire, leaving a web of fibres. The pulp sheet is passed through a series of rotating rolls ("presses") that squeeze out water and air until the fibre consistency is 40 to 45%. The sheet is then floated through a multi-storey sequence of hot-air dryers until the consistency is 90 to 95%. Finally, the continuous pulp sheet is cut into pieces and stacked into bales. The pulp bales are compressed, wrapped and packaged into bundles for storage and transport.

Fig. Wet end of pulp machine showing fibre mat on the wire

Although similar in principle to making pulp sheets, paper making is considerably more complex. Some mills use a variety of different pulps to optimize paper quality (e.g., a mix of hardwood, softwood, kraft, sulphite, mechanical or recycled pulps). Depending on the type of pulp used, a series of steps is necessary prior to forming the paper sheet. Generally, dried market pulp is rehydrated, while high-consistency pulp from storage is diluted. Pulp fibres may be beaten to increase the fibre-bonding area and thereby improve paper sheet strength. The pulp is then blended with "wet-end" additives and passed through a final set of screens and cleaners. The pulp is then ready for the paper machine.

Table : Papermaking additives

Additive	Location applied	Purpose and/or examples of specific agents
Most commonly used additives		
Talc	Wet end	Pitch control (prevent deposition and accumulation of pitch) Filler (make brighter, smoother, more opaque)
Titanium dioxide	Wet end	Pigment (brighten sheet, improve printing) Filler (make brighter, smoother, more opaque)
"Alum"($Al_2(SO_4)_3$)	Wet end	Precipitates rosin sizing onto fibres Retention aid (fix additives to fibres, improve pulp fibre retention)
Rosin	Wet end	Internal sizing (resist liquid penetration)
Clay (kaolin)	Wet/dry	Filler (make brighter, smoother, more opaque) Pigment or surface coating (impart colour)
Starch	Wet/dry	Surface sizing (resist liquid penetration) Dry strength additive (increase strength, reduce surface lint) Retention aid (bind additives to paper, improve pulp fibre retention)
Dyes and pigments	Wet/dry	e.g., acid, basic or direct dyes,colour lakes, $CaCO_3$, may also include solvent vehicles
Latex	Dry end	Adhesive (reinforce sheet, bind additives to paper, fill pores) Waterproofing (resist liquid penetration)
Other additives		
Slimicides	Wet end	e.g., thiones, thiazoles, thiocyanates, hiocarbamates, thiols, isothiazolinones, formaldehyde, glutaraldehyde, glycols, naphthol, chlorinated and brominated organics, organic mercury compounds
Defoamers	Wet end	e.g., pine oil, fuel oil, recycled oils, silicones, alcohols
Wire treatment agents	Wet end	e.g., imidazoles, butyl diglycol, acetone, turpentine, phosphoric acid
Wet and dry strength additives	Wet end	e.g., formaldehyde resins, epichlorohydrin, glyoxal, gums, polyamines, phenolics, polyacrylamides, polyamids, cellulose derivatives
Coatings, adhesives and plasticizers	Dry end	e.g., aluminium hydroxide, polyvinyl acetate, acrylics, linseed oil, gums, protein glues, wax emulsions, azite, glyoxal, stearates, solvents, polyethylene, cellulose derivatives, foil, rubber derivatives, polyamines, polyesters, butadiene-styrene polymers
Others	Wet/dry	Corrosion inhibitors, dispersants, flameproofing, antitarnish agents, drainage aids, deflocculants, pH control agents, preservatives

The flow spreader and headbox distribute a thin suspension (1 to 3%) of refined pulp onto a moving wire (similar to a pulp machine, only at a much higher speed, sometimes in excess of 55 km/h) which forms the fibres into a thin felted sheet. The sheet moves through a series of press rolls to the dryer section, where a series of steam-heated rolls evaporate most of the remaining water. Hydrogen bonds between the fibres have fully developed at this stage. Finally, the paper is calendered and reeled. Calendering is the process by which the paper surface is ironed smooth and its thickness reduced. The dried, calendered paper sheet is wound onto a reel, labelled and transported to the warehouse. "Dry-end" additives can be added before calendering on the paper machine or in separate "off-machine" coating operations in the converting sector of the industry.

Fig. Dry end of a paper machine showing full paper reel and operator using air slitter to cut end

A variety of chemicals are used in the papermaking process to provide the paper with specific surface characteristics and sheet properties. The most commonly used additives are typically used at the per cent level, though some such as clay and talc may contribute as much as 40% to the dry weight of certain papers. The diversity of chemical additives which may be used for specific production purposes and products; some of these are used at very low concentrations (e.g., slimicides are added to process water in parts per million).

The process of making paperboard is similar to that of making paper or pulp. A suspension of pulp and water is dispersed onto a travelling wire, the water is removed, and the sheet dried and stored as a roll. The process differs in the way that the sheet is formed to give thickness, in the combining of multiple layers, and in the drying process. Board can be made from single or multi-layered sheets with or without a core. The sheets are usually high-quality kraft pulp (or kraft and CTMP blend), while the core is made from either a blend of semi-chemical and low-cost recycled pulp or from entirely recycled pulp and other waste material. Coatings, vapour barriers and multiple layers are added according to the end use to protect the contents from water and physical damage.

POWER GENERATION AND WATER TREATMENT

In addition to liquor recovery, pulp mills recover a significant portion of energy from burning waste materials and by-products of the process in power boilers. Materials such as bark, wood waste and dried sludge collected from effluent treatment systems may be burned to provide steam to power electrical generators.

Pulp and paper mills consume vast amounts of fresh water. A 1,000 tonne per day bleached kraft pulp mill may use more than 150 million litres of water a day; a paper mill even more. In order to prevent adverse effects on mill equipment and to maintain product quality, the incoming water must be treated to remove contaminants, bacteria and minerals. Several treatments are applied depending on the quality of the incoming water. Sedimentation beds, filters, flocculants, chlorine and ion exchange resins are all used to treat water before it is used in the process. Water that is used in the power and recovery boilers is further treated with oxygen scavengers and corrosion inhibitors such as hydrazine and morpholine to avoid deposits forming in the boiler tubes, to reduce metal corrosion, and to prevent carry-over of water to the steam turbine.

CHEMICAL AND BY-PRODUCT PRODUCTION

Because many bleaching chemicals are reactive and hazardous to transport, they are produced on-site or nearby. Chlorine dioxide (ClO_2), sodium hypochlorite (NaOCl) and peracids are always produced on-site, while chlorine (Cl_2) and sodium hydroxide or caustic (NaOH) are usually produced off-site. Tall oil, a product derived from the resin and fatty acids that are extracted during kraft cooking, may be refined on- or off-site. Turpentine, a lighter fraction kraft by-product, is often collected and concentrated on-site, and refined elsewhere.

CHLORINE DIOXIDE

Chlorine dioxide (ClO_2) is a highly reactive greenish-yellow gas. It is toxic and corrosive, explodes at high concentrations (10%) and is quickly reduced to Cl_2 and O_2 in the presence of ultraviolet light. It must be prepared as a dilute gas and stored as a dilute liquid, making bulk transport impossible.

ClO_2 is generated by reducing sodium chlorate (Na_2ClO_3) with either SO_2, methanol, salt or hydrochloric acid. The gas leaving the reactor is condensed and stored as a 10% liquid solution. Modern ClO_2generators operate at 95% efficiency or greater, and the small amount of Cl_2 that is produced will be collected or scrubbed out of the vent gas. Side reactions may occur depending on the purity of the feed chemicals, the temperature and other process variables. By-products are returned to the process and spent chemicals are neutralized and sewered.

SODIUM HYPOCHLORITE

Sodium hypochlorite (NaOCl) is produced by combining Cl_2 with a dilute solution of NaOH. It is a simple, automated process that requires almost no intervention. The process is controlled by maintaining the caustic concentration such that the residual Cl_2 in the process vessel is minimized.

CHLORINE AND CAUSTIC

Chlorine (Cl_2), used as a bleaching agent since the early 1800s, is a highly reactive, toxic, green-coloured gas which becomes corrosive when moisture is present. Chlorine is usually manufactured by the electrolysis of brine (NaCl) into Cl_2 and NaOH at regional installations, and transported to the customer as a pure liquid. Three methods are used to produce Cl_2 on an industrial scale: the mercury cell, the diaphragm cell, and the most recent development, the membrane cell. Cl_2 is always produced at the anode. It is then cooled, purified, dried, liquefied and transported to the mill. At large or remote pulp mills, local facilities may be constructed, and the Cl_2 can be transported as a gas.

The quality of NaOH depends on which of the three processes is used. In the older mercury cell method, the sodium and mercury combine to form an amalgam that is decomposed with water. The resulting NaOH is nearly pure. One of the shortcomings of this process is that mercury contaminates the workplace and has resulted in serious environmental problems. The NaOH produced from the diaphragm cell is removed with the spent brine and concentrated to allow the salt to crystallize and separate. Asbestos is used as the diaphragm. The purest NaOH is produced in membrane cells. A semi-permeable resin-based membrane allows sodium ions to pass through without the brine or chlorine ions, and combine with water added to the cathode chamber to form pure NaOH. Hydrogen gas is a by-product of each process. It is usually treated and used either in other processes or as fuel.

TALL OIL PRODUCTION

Kraft pulping of highly-resinous species such as pine produces sodium soaps of resin and fatty acids. The soap is collected from black liquor storage tanks and from soap skimming tanks that are located in the evaporator train of the chemical recovery process. Refined soap or tall oil can be used as a fuel additive, dust control agent, road stabilizer, pavement binder and roofing flux.

At the processing plant, soap is stored in primary tanks to allow the black liquor to settle to the bottom. The soap rises and overflows into a second storage tank. Sulphuric acid and the decanted soap are fed into a reactor, heated to 100°C, agitated and then allowed to settle. After settling overnight, the crude tall oil is decanted into a storage vessel and allowed to sit for another day. The top fraction is considered dry crude tall oil and is pumped to storage, ready for shipment. The cooked lignin in the bottom fraction will become part of the subsequent batch. The spent sulphuric acid is pumped to a storage tank, and any entrained lignin is allowed to settle to the bottom. The lignin left in the reactor is concentrated for several cooks, dissolved in 20% caustic and returned to the primary soap tank. Periodically, the collected black liquor and the residual lignin from all sources are concentrated and burned as fuel.

TURPENTINE RECOVERY

Gases from the digesters and condensate from black liquor evaporators may be collected for recovery of turpentine. The gases are condensed, combined, then stripped of turpentine, which is recondensed, collected and sent to a decanter. The top fraction of the decanter is drawn off and sent to storage, while the bottom fraction is recycled to the stripper. Raw turpentine is stored separately from the rest of the collection system because it is noxious and flammable, and is usually processed off-site. All the non-condensable gases are collected and incinerated either in the power boilers, the lime kiln or a dedicated furnace. The turpentine can be processed for use in camphor, synthetic resins, solvents, flotation agents and insecticides.

OCCUPATIONAL HAZARDS AND CONTROLS

An overview of the types of exposures which may be expected in each area of pulp and paper operations. Although exposures may be listed as specific to certain production processes, exposures to employees from other areas may also occur depending on weather conditions, proximity to sources of exposure, and whether they work in more than one process area (e.g., quality control, general labour pool and maintenance personnel).

RMP = refining mechanical pulping; CMP = chemi-mechanical pulping; CTMP = chemi-thermomechanical pulping.

Exposure to the potential hazards is likely to depend on the extent of automation of the plant. Historically, industrial pulp and paper production was a semi-automatic process which required a great deal of manual intervention. In such facilities, operators would sit at open panels adjacent to the processes to view the effects of their actions. The valves at the top and bottom of a batch digester would be manually opened, and during the filling stages, gases in the digester would be displaced by the incoming chips. Chemical levels would be adjusted based on experience rather than sampling, and process adjustments would be dependent on the skill and knowledge of the operator, which at times led to upsets. For example, over-chlorination of pulp would expose workers downstream to increased levels of bleaching agents. In most modern mills, progress from manually controlled to electronically controlled pumps and valves allows for remote operation. The demand for process control within narrow tolerances has required computers and sophisticated engineering strategies. Separate control rooms are used to isolate the electronic equipment from the pulp and paper production environment. Consequently, operators usually work in air-conditioned control rooms which offer refuge from the noise, vibration, temperature, humidity and chemical exposures inherent to mill operations.

Fig. Worker opening cap on manually controlled batch digester

Safety hazards including nip points, wet walking surfaces, moving equipment and heights are common throughout pulp and paper operations. Guards around moving conveyors and machinery parts, quick clean-up of spills, walking surfaces which allow drainage, and guard-rails on walkways adjacent to production lines or at height are all essential. Lock-out procedures must be followed for maintenance of chip conveyors, paper machine rolls and all other machinery with moving parts. Mobile equipment used in chip storage, dock and shipping areas, warehousing and other operations should have roll-over protection, good visibility and horns; traffic lanes for vehicles and pedestrians should be clearly marked and signed.

Noise and heat are also ubiquitous hazards. The major engineering control is operator enclosures, as described above, usually available in wood preparation, pulping, bleaching and sheet-forming areas. Air-conditioned enclosed cabs for mobile equipment used in chip pile and other yard operations are also available. Outside these enclosures, workers usually require hearing protection. Work in hot process or outdoor areas and in vessel maintenance operations requires workers to be trained to recognize symptoms of heat stress; in such areas, work scheduling should allow acclimatization and rest periods. Cold weather may create frostbite hazards in outdoor jobs, as well as foggy conditions near chip piles, which remain warm.

Wood, its extracts and associated micro-organisms are specific to wood preparation operations and the initial stages of pulping. Control of exposures will depend on the particular operation, and may include operator booths, enclosure and ventilation of saws and conveyors, as well as enclosed chip storage and low chip inventory. Use of compressed air to clear wood dust creates high exposures and should be avoided.

Chemical pulping operations present the opportunity for exposures to digestion chemicals as well as gaseous by-products of the cooking process, including reduced (kraft pulping) and oxidized (sulphite pulping) sulphur compounds and volatile organics. Gas formation may be influenced by a number of operating conditions: the wood species used; the quantity of wood pulped; the amount and concentration of white liquor applied; the amount of time required for pulping; and maximum temperature attained. In addition to automatic digester capping valves and operator control rooms, other controls for these areas include local exhaust ventilation at batch digesters and blow tanks, capable of venting at the rate the vessel's gases are released; negative pressure in recovery boilers and sulphite-SO_2 acid towers to prevent gas leaks; ventilated full or partial enclosures over post-digestion washers; continuous gas monitors with alarms where leaks may occur; and emergency response planning and training. Operators taking samples and conducting tests should be aware of the potential for acid and caustic exposure in process and waste streams, and the possibility of side reactions such as hydrogen sulphide gas (H_2S) production if black liquor from kraft pulping comes into contact with acids (e.g., in sewers).

In chemical recovery areas, acidic and alkaline process chemicals and their by-products may be present at temperatures in excess of 800°C. Job responsibilities may require workers to come into direct contact with these chemicals, making heavy duty clothing a necessity. For example, workers rake the spattering molten smelt that collects at the base of the boilers, thereby risking chemical and thermal burns. Workers may be exposed to dust when sodium sulphate is added to concentrated black liquor, and any leak or opening will release noxious (and potentially fatal) reduced sulphur gases. The potential for a smelt water explosion always exists around the recovery boiler. Water leaks in the tube walls of the boiler have resulted in several fatal explosions. Recovery boilers should be shut down at any indication of a leak, and special procedures should be implemented for transferring the smelt. Loading of lime and other caustic materials should be done with enclosed and ventilated conveyors, elevators and storage bins.

In bleach plants, field operators may be exposed to the bleaching agents as well as chlorinated organics and other by-products. Process variables such as bleaching chemical strength, lignin content, temperature and pulp consistency are constantly monitored, with operators collecting samples and performing laboratory tests. Because of the hazards of many of the bleaching agents used, continuous alarm monitors should be in place, escape respirators should be issued to all employees, and operators should be trained in emergency response procedures. Canopy enclosures with dedicated exhaust ventilation are standard engineering controls found at the top of each bleaching tower and washing stage.

Chemical exposures in the machine room of a pulp or paper mill include chemical carry-over from the bleach plant, the papermaking additives and the chemical mixture in the waste water. Dusts (cellulose, fillers, coatings) and exhaust fumes from mobile equipment are present in the dry-end and the finishing operations. Cleaning between product runs may be done with solvents, acids and alkalis. Controls in this area may include complete enclosure over the sheet drier; ventilated enclosure of the areas where additives are unloaded, weighed and mixed; use of additives in liquid rather than powder form; use of water-based rather than solvent-based inks and dyes; and eliminating the use of compressed air to clean up trimmed and waste paper.

Paper production in recycled paper plants is generally dustier than conventional paper production using newly produced pulp. Exposure to micro-organisms can occur from the beginning (paper collection and separation) to the end (paper production) of the production chain, but exposure to chemicals is less important than in conventional paper production.

Pulp and paper mills employ an extensive maintenance group to service their process equipment, including carpenters, electricians, instrument mechanics, insulators, machinists, masons, mechanics, millwrights, painters, pipefitters, refrigeration mechanics, tinsmiths and welders. Along with their trade-specific exposures, these tradespeople may be exposed to any of the process-related hazards. As mill operations have become more automated and enclosed, the maintenance, cleaning and quality assurance operations have become the most highly exposed. Plant shutdowns to clean vessels and machines are of special concern. Depending on mill organization, these operations may be carried out by in-house maintenance or production personnel, although subcontracting to non-mill personnel, who may have less occupational health and safety support services, is common.

In addition to process exposures, pulp and paper mill operations entail some noteworthy exposures for maintenance personnel. Because pulping, recovery and boiler operations involve high heat, asbestos was used extensively to insulate pipes and vessels. Stainless steel is often used in vessels and pipes throughout pulping, recovery and bleaching operations, and to some extent in papermaking. Welding this metal is known to generate chromium and nickel fumes. During maintenance shut-downs, chromium-based sprays may be applied to protect the floor and walls of recovery boilers from corrosion during start-up operations. Process quality measurements in the production line are often made using infrared and radio-isotope gauges. Although the gauges are usually well shielded, instrument mechanics who service them may be exposed to radiation.

Some special exposures may also occur among employees in other mill-support operations. Power boiler workers handle bark, waste wood and sludge from the effluent treatment system. In older mills, workers remove ash from the bottom of the boilers and then reseal the boilers by applying a mixture of asbestos and cement around the boiler grate. In modern power boilers, this process is automated. When material is fed into the boiler at too high a moisture level, workers may be exposed to blow-backs of incomplete combustion products. Workers responsible for water treatment may be exposed to chemicals such as chlorine, hydrazine and various resins. Because of the reactivity of ClO_2, the ClO_2 generator is usually located in a restricted area and the operator is stationed in a remote control room with excursions to collect samples and service the saltcake filter. Sodium chlorate (a strong oxidizer) used to generate ClO_2 can become dangerously flammable if it is allowed to spill on any organic or combustible material and then dry. All spills should be wetted down before any maintenance work may proceed, and all equipment should be thoroughly cleaned afterward. Wet clothing should be kept wet and separate from street clothing, until washed.

INJURIES AND NON-MALIGNANT DISEASES

INJURIES

Only limited statistics are available on accident rates in general in this industry. Compared to other manufacturing industries, the 1990 accident rate in Finland was below the average; in Canada, the rates from 1990 to 1994 were similar to other industries; in the United States, the 1988 rate was slightly above average; in Sweden and Germany, the rates were 25% and 70% above the average (ILO 1992; Workers' Compensation Board of British Columbia 1995).

The most commonly encountered risk factors for serious and fatal accidents in the pulp and paper industry are the papermaking equipment itself and the extreme size and weight of pulp or paper bales and rolls. In a 1993 United States government study of occupational fatalities from 1979 to 1984 in pulp, paper and paperboard mills (US Department of Commerce 1993), 28% were due to workers being caught in or between rotating rolls or equipment ("nip-points") and 18% were due to workers being crushed by falling or tumbling objects, especially rolls and bales. Other causes of multiple deaths included electrocution, hydrogen sulphide and other toxic gas inhalation, massive thermal/chemical burns and one case of heat exhaustion. The number of serious accidents associated with paper machines has been reported to decrease with the installation of newer equipment in some countries. In the converting sector, repetitive and monotonous work, and the use of mechanized equipment with higher speeds and forces, has become more common. Although no sector-specific data are available, it is expected that this sector will experience greater rates of over-exertion injuries associated with repetitive work.

NON-MALIGNANT DISEASES

The most well documented health problems encountered by pulp mill workers are acute and chronic respiratory disorders (Torén, Hagberg and Westberg 1996). Exposure to extremely high concentrations of chlorine, chlorine dioxide or sulphur dioxide may occur as a result of a leak or other process upset. Exposed workers may develop acute chemical-induced lung injury with severe inflammation of air passages and release of fluid into the air spaces, requiring hospitalization. The extent of damage depends on the duration and intensity of the exposure, and the specific gas involved. If the worker survives the acute episode, complete recovery may occur. However, in less intense exposure incidents (also usually as a result of process upsets or spills), acute exposure to chlorine or chlorine dioxide may trigger the subsequent development of asthma.

This irritant-induced asthma has been recorded in numerous case reports and recent epidemiological studies, and current evidence indicates that it may persist for many years following the exposure incident. Workers similarly exposed who do not develop asthma may experience persistently increased nasal irritation, cough, wheezing and reduction in airflow rates. Workers most at risk for these exposure incidents include maintenance workers, bleach plant workers and construction workers at pulp mill sites. High levels of chlorine dioxide exposure also cause eye irritation and the sensation of seeing halos around lights.

Some mortality studies have indicated increased risk of death from respiratory disease among pulp mill workers exposed to sulphur dioxide and paper dust (Jäppinen and Tola 1990; Torén, Järvholm and Morgan 1989). Increased respiratory symptoms have also been reported in sulphite mill workers who are chronically exposed to low levels of sulphur dioxide (Skalpe 1964), although increased airflow obstruction is not normally reported among pulp mill populations in general. Symptoms of respiratory irritation are also reported by workers exposed to high air concentrations of terpenes in turpentine recovery processes often present at pulp mill sites. Soft paper dust has also been reported to be associated with increased asthma and chronic obstructive pulmonary disease (Torén, Hagberg and Westberg 1996).

Exposure to micro-organisms, especially around wood chip and waste piles, debarkers and sludge presses, creates an increased risk for hypersensitivity responses in the lungs. Evidence for this appears to be limited to isolated case reports of hypersensitivity pneumonitis, which can lead to chronic lung scarring.

Bagassosis, or hypersensitivity pneumonitis associated with exposure to thermophylic micro-organisms and bagasse (a sugar cane by-product), is still seen in mills using bagasse for fibre.

Other respiratory hazards commonly encountered in the pulp and paper industry include stainless steel welding fumes and asbestos. Maintenance workers are the group most likely to be at risk from these exposures.

Reduced sulphur compounds (including hydrogen sulphide, dimethyl disulphides and mercaptans) are potent eye irritants and may cause headaches and nausea in some workers.

These compounds have very low odour thresholds (ppb range) in individuals not previously exposed; however, among long-time workers in the industry, odour thresholds are considerably higher. Concentrations in the range of 50 to 200 ppm produce olfactory fatigue, and subjects can no longer detect the distinctive "rotten eggs" odour. At higher concentrations, exposure will result in unconsciousness, respiratory paralysis and death. Fatalities associated with exposure to reduced sulphur compounds in confined spaces have occurred at pulp mill sites.

Cardiovascular mortality has been reported to be increased in pulp and paper workers, with some exposure-response evidence suggesting a possible link with exposure to reduced sulphur compounds (Jäppinen 1987; Jäppinen and Tola 1990). However, other causes for this increased mortality may include noise exposure and shift work, both of which have been associated with increased risk for ischaemic heart disease in other industries.

Skin problems encountered by pulp and paper mill workers include acute chemical and thermal burns and contact dermatitis (both irritant and allergic). Pulp mill workers in kraft process mills frequently experience alkali burns to the skin as a result of contact with hot pulping liquors and calcium hydroxide slurries from the recovery process. Contact dermatitis is reported more frequently among paper mill and converting workers, as many of the additives, defoaming agents, biocides, inks and glues used in paper and paper-product making are primary skin irritants and sensitizers. Dermatitis may occur from exposure to the chemicals themselves or from handling freshly treated paper or paper products.

Noise is a significant hazard throughout the pulp and paper industry. The US Department of Labor estimated that noise levels over 85 dBA were found in over 75% of plants in the paper and allied products industries, compared to 49% of plants in manufacturing in general, and that over 40% of workers were exposed regularly to noise levels over 85 dBA (US Department of Commerce 1983). Noise levels around paper machines, chippers and recovery boilers tend to be well over 90 dBA. Conversion operations also tend to generate high noise levels. Reduction in worker exposure around paper machines is usually attempted by the use of enclosed control rooms. In converting, where the operator is usually stationed next to the machine, this type of control measure is seldom used. However where converting machines have been enclosed, this has resulted in decreased exposure to both paper dust and noise.

Excessive heat exposure is encountered by paper mill workers working in paper machine areas, with temperatures of 60°C being recorded, although no studies of the effects of heat exposure in this population are available in the published scientific literature.

CANCER

Table : Summary of studies on lung cancer, stomach cancer, lymphoma and leukaemia in pulp and paper workers

Process description	Location of study	Type of study	Lung cancer	Stomach cancer	Lymphoma NHL/HD§	Leukaemia
Sulphite	Finland	C	0.9	1.3	X/X	X
Sulphite	USA	C	1.1	0.7	—	0.9
Sulphite	USA	C	0.8	1.5	1.3/X	0.7
Sulphite	USA	PM	0.9	2.2*	2.7*/X	1.3
Sulphate	Finland	C	0.9	0.9	0/0	X
Sulphate	USA	C	0.8	1.0	2.1/0	0.2
Sulphate	USA	PM	1.1	1.9	1.1/4.1*	1.7
Chlorine	Finland	C	3.0*	—	—	—
Sulphite/paper	Sweden	CR	—	2.8*	—	—
Paper dust	Canada	CR	2.0*	—	—	—
Paper mill	Finland	C	2.0*	1.7	X/X	—
Paper mill	Sweden	C	0.7*	—	—	—
Paper mill	USA	C	0.8	2.0	—	2.4
Paper mill	Sweden	CR	1.6	—	—	—
Paper mill	USA	PM	1.3	0.9	X/1.4	1.4
Board mill	Finland	C	2.2*	0.6	X/X	X
Power plant	Finland	C	0.5	2.1	—	—
Maintenance	Finland	C	1.3	0.3*	1.0/X	1.5
Maintenance	Sweden	CR	2.1*	0.8	—	—
Pulp and paper	USA	C	0.9	1.2	0.7/X	1.8
Pulp and paper	USA	C	0.8	1.2	1.7/X	0.5
Pulp and paper	Sweden	CR	0.8	1.3	1.8	1.1
Pulp and paper	Sweden	CR	—	—	2.2/0	—
Pulp and paper	Sweden	CR	1.1	0.6	—	—
Pulp and paper	USA	CR	1.2*	—	—	—
Pulp and paper	USA	CR	1.1	—	—	—
Pulp and paper	USA	CR	—	—	—/4.0	—
Pulp and paper	Canada	PM	—	1.2	3.8*/—	—
Pulp and paper	USA	PM	1.5*	0.5	4.4/4.5	2.3
Pulp and paper	USA	PM	0.9	1.7*	1.6/1.0	1.1
Pulp and paper	USA	PM	0.9	1.2	1.5/1.9*	1.4
Pulp and paper	USA	PM	—	1.7*	1.4	1.6*

C = cohort study, CR = case-referent study, PM = proportionate mortality study. * Statistically significant. = Where separately reported, NHL = non Hodgkin lymphoma and HD = Hodgkin's disease. X = 0 or 1 case reported, no risk estimate calculated, — = No data reported.

A risk estimate exceeding 1.0 means the risk is increased, and a risk estimate below 1.0 indicates decreased risk.

Exposures to numerous substances designated by the International Agency for Research on Cancer (IARC) as known, probable and possible carcinogens may occur in pulp and paper operations. Asbestos, known to cause lung cancer and mesothelioma, is used to insulate pipes and boilers. Talc is used extensively as a paper additive, and can be contaminated with asbestos. Other paper additives, including benzidine-based dyes, formaldehyde and epichlorohydrin, are considered probable human carcinogens. Hexavalent chromium and nickel compounds, generated in stainless-steel welding, are known lung and nasal carcinogens. Wood dust has recently been classified by IARC as a known carcinogen, based mainly on evidence of nasal cancer among workers exposed to hardwood dust (IARC, 1995). Diesel exhaust, hydrazine, styrene, mineral oils, chlorinated phenols and dioxins, and ionizing radiation are other probable or possible carcinogens which may be present in mill operations.

Few epidemiological studies specific to pulp and paper operations have been conducted, and they indicate few consistent results. Exposure classifications in these studies have often used the broad industrial category "pulp and paper", and even the most specific classifications grouped workers by types of pulping or large mill areas. The three cohort studies in the literature to date involved fewer than 4,000 workers each. Several large cohort studies are currently under way, and IARC is coordinating an international multicentric study likely to include data from more than 150,000 pulp and paper workers, allowing much more specific exposure analyses.

RESPIRATORY SYSTEM CANCERS

Maintenance workers in paper and pulp mills experience an increased risk of lung cancer and malignant mesotheliomas, probably because of their exposure to asbestos. A Swedish study showed a threefold increased risk of pleural mesothelioma among pulp and paper workers (Malker et al. 1985). When the exposure was further analysed, 71% of the cases had been exposed to asbestos, the majority having worked in mill maintenance. Elevations in lung cancer risk among maintenance workers have also been shown in Swedish and Finnish pulp and paper mills (Torén, Sällsten and Järvholm 1991; Jäppinen et al. 1987).

In the same Finnish study, a twofold increased risk of lung cancer was also observed among both paper mill and board mill workers. The investigators made a subsequent study restricted to pulp mill workers exposed to chlorine compounds, and found a threefold increased risk of lung cancer.

Few other studies of pulp and paper workers have shown increased risks for lung cancer. A Canadian study showed an increased risk among those exposed to paper dust (Siemiatycki et al. 1986), and US and Swedish studies showed increased risks among paper mill workers (Milham and Demers 1984; Torén, Järvholm and Morgan 1989).

GASTRO-INTESTINAL CANCERS

Increased risk of stomach cancer has been indicated in many studies, but the risks are not clearly associated with any one area; therefore the relevant exposure is unknown. Socio-economic status and dietary habits are also risk factors for stomach cancer, and might be confounders; these factors were not taken into account in any of the studies reviewed.

The association between gastric cancer and pulp and paper work was first seen in a US study in the 1970s (Milham and Demers 1984). The risk was found to be even higher, nearly doubled, when sulphite workers were examined separately. US sulphite and groundwood workers were also found in a later study to run an increased risk of stomach cancer (Robinson, Waxweiller and Fowler 1986). A risk of the same magnitude was found in a Swedish study among pulp and paper mill workers from an area where only sulphite pulp was produced (Wingren et al. 1991). American paper, paperboard and pulp mill workers in New Hampshire and Washington state ran an increased mortality from stomach cancer (Schwartz 1988; Milham 1976). The subjects were probably a mixture of sulphite, sulphate and paper mill workers. In a Swedish study, threefold increased mortality due to stomach cancer was found in a group comprising sulphite and paper mill workers (Wingren, Kling and Axelson 1985). The majority of pulp and paper studies reported excesses of stomach cancer, though some did not.

Due to the small number of cases, most studies of other gastrointestinal cancers are inconclusive. An increased risk of colon cancer among workers in the sulphate process and in paper board production has been reported in a Finnish study (Jäppinen et al. 1987), as well as among US pulp and paper workers (Solet et al. 1989). The incidence of biliary tract cancer in Sweden between 1961 and 1979 was linked with occupational data from the 1960 National Census (Malker et al. 1986). An increased incidence of cancer of the gallbladder among male paper mill workers was identified. Increased risks of pancreatic cancer have been observed in some studies of paper mill workers and sulphite workers (Milham and Demers 1984; Henneberger, Ferris and Monson 1989), as well as in the broad group of pulp and paper workers (Pickle and Gottlieb 1980; Wingren et al. 1991). These findings have not been substantiated in other studies.

HAEMATOLOGICAL MALIGNANCIES

The issue of lymphomas among pulp and paper mill workers was originally addressed in a US study from the 1960s, where a fourfold increased risk of Hodgkin's disease was found among pulp and paper workers (Milham and Hesser 1967). In a subsequent study, the mortality among pulp and paper mill workers in the state of Washington between 1950 and 1971 was investigated, and a doubled risk of both Hodgkin's disease and multiple myeloma was observed (Milham 1976). This study was followed by one analysing mortality among pulp and paper union members in the United States and Canada (Milham and Demers 1984). It showed almost a threefold increased risk for lymphosarcoma and reticulum cell sarcoma among sulphite workers, while sulphate workers had a fourfold increased risk of Hodgkin's disease. In a US cohort study, sulphate workers were observed to have a twofold risk of lymphosarcoma and reticulosarcoma (Robinson, Waxweiller and Fowler 1986).

In many of the studies where it was possible to investigate the occurrence of malignant lymphomas, an increased risk has been found (Wingren et al. 1991; Persson et al. 1993). Since the increased risk occurs both in sulphate and sulphite mill workers, this points towards a common source of exposure. In the sorting and chipping departments, the exposures are rather similar. The workforce is exposed to wood dust, terpenes and other extractable compounds from the wood. In addition, both pulping processes bleach with chlorine, which has the potential to create chlorinated organic by-products, including small amounts of dioxins.

Compared with lymphomas, studies on leukaemias show less consistent patterns, and the risk estimates are lower.

OTHER MALIGNANCIES

Among US paper mill workers with presumed exposure to formaldehyde, four cases of urinary tract cancer were found after 30 years' latency, although only one was expected (Robinson, Waxweiller and Fowler 1986). All of these individuals had worked in the paper-drying areas of the paper mills.

In a case-control study from Massachusetts, central nervous system tumours in childhood were associated with an unspecified paternal occupation as a paper and pulp mill worker (Kwa and Fine 1980). The authors regarded their observation as a random event. However, in three subsequent studies, increased risks were also found (Johnson et al. 1987; Nasca et al. 1988; Kuijten, Bunin and Nass 1992). In studies from Sweden and Finland, two- to threefold increased risks of brain tumours were observed among pulp and paper mill workers.

ENVIRONMENTAL AND PUBLIC HEALTH ISSUES

Because the pulp and paper industry is a large consumer of natural resources (i.e., wood, water and energy), it can be a major contributor to water, air and soil pollution problems and has come under a great deal of scrutiny in recent years. This concern appears to be warranted, considering the quantity of water pollutants generated per tonne of pulp (e.g., 55 kg of biological oxygen demand, 70 kg of suspended solids, and up to 8 kg of organochlorine compounds) and the amount of pulp produced globally on an annual basis (approximately 180 million tonnes in 1994). In addition, only about 35% of used paper is recycled, and waste paper is a major contributor to total worldwide solid waste (about 150 million of 500 million tonnes annually).

Historically, pollution control was not considered in the design of pulp and paper mills. Many of the processes used in the industry were developed with little regard for minimizing effluent volume and pollutant concentration. Since the 1970s, pollution abatement technologies have become integral components of mill design in Europe, North America and other parts of the world.

AIR POLLUTION ISSUES

Air emissions of oxidized sulphur compounds from pulp and paper mills have caused damage to vegetation, and emissions of reduced sulphur compounds have generated complaints about "rotten egg" odours. Studies among residents of pulp mill communities, in particular children, have shown respiratory effects related to particulate emissions, and mucous membrane irritation and headache thought to be related to reduced sulphur compounds. Of the pulping processes, those with the greatest potential to cause air pollution problems are chemical methods, in particular kraft pulping.

Sulphur oxides are emitted at the highest rates from sulphite operations, especially those using calcium or magnesium bases. The major sources include batch digester blows, evaporators and liquor preparation, with washing, screening and recovery operations contributing lesser amounts. Kraft recovery furnaces are also a source of sulphur dioxide, as are power boilers which use high-sulphur coal or oil as fuel.

Reduced sulphur compounds, including hydrogen sulphide, methyl mercaptan, dimethyl sulphide and dimethyl disulphide, are almost exclusively associated with kraft pulping, and give these mills their characteristic odour. The major sources include the recovery furnace, digester blow, digester relief valves, and washer vents, though evaporators, smelt tanks, slakers, the lime kiln and waste water may also contribute. Some sulphite operations use reducing environments in their recovery furnaces and may have associated reduced sulphur odour problems.

Sulphur gases emitted by the recovery boiler are best controlled by reducing emissions at the source. Controls include black liquor oxidation, reduction in liquor sulphidity, low-odour recovery boilers and proper operation of the recovery furnace. Sulphur gases from digester blow, digester relief valves and liquor evaporation can be collected and incinerated - for example, in the lime kiln. Combustion flue gases can be collected using scrubbers.

Nitrogen oxides are produced as products of high-temperature combustion, and may arise in any mill with a recovery boiler, power boiler or lime kiln, depending on the operating conditions. The formation of nitrogen oxides can be controlled by regulating temperatures, air-fuel ratios and residence time in the combustion zone. Other gaseous compounds are minor contributors to mill air pollution (e.g., carbon monoxide from incomplete combustion, chloroform from bleaching operations, and volatile organics from digester relief and liquor evaporation).

Particulates arise mainly from combustion operations, though smelt-dissolving tanks can also be a minor source. More than 50% of pulp mill particulate is very fine (less than 1 mm in diameter). This fine material includes sodium sulphate (Na_2SO_4) and sodium carbonate (Na_2CO_3) from recovery furnaces, lime kilns and smelt-dissolving tanks, and NaCl from burning by-products of logs which have been stored in salt water. Lime kiln emissions include a significant amount of coarse particulates due to entrainment of calcium salts and sublimation of sodium compounds. Coarse particulate may also include fly ash and organic combustion products, especially from power boilers. Reduction of particulate concentrations can be achieved by passing flue gases through electrostatic precipitators or scrubbers. Recent innovations in power boiler technology include fluidized bed incinerators which burn at very high temperatures, result in more efficient energy conversion, and allow burning of less uniform wood waste.

WATER POLLUTION ISSUES

Contaminated wastewater from pulp and paper mills can cause death of aquatic organisms, allow bioaccumulation of toxic compounds in fish, and impair the taste of downstream drinking water. Pulp and paper wastewater effluents are characterized on the basis of physical, chemical or biological characteristics, with the most important being solids content, oxygen demand and toxicity.

The solids content of wastewater is typically classified on the basis of the fraction that is suspended (versus dissolved), the fraction of suspended solids that is settleable, and the fractions of either that are volatile. The settleable fraction is the most objectionable because it may form a dense sludge blanket close to the discharge point, which rapidly depletes dissolved oxygen in the receiving water and allows the proliferation of anaerobic bacteria which generate methane and reduced sulphur gases. Although non-settleable solids are usually diluted by the receiving water and are therefore of less concern, they may transport toxic organic compounds to aquatic organisms. Suspended solids discharged from pulp and paper mills include bark particles, wood fibre, sand, grit from mechanical pulp grinders, papermaking additives, liquor dregs, by-products of water treatment processes and microbial cells from secondary treatment operations.

Wood derivatives dissolved in the pulping liquors, including oligosaccharides, simple sugars, low-molecular-weight lignin derivatives, acetic acid and solubilized cellulose fibres, are the main contributors to both biological oxygen demand (BOD) and chemical oxygen demand (COD). Compounds which are toxic to aquatic organisms include chlorinated organics (AOX; from bleaching, especially kraft pulp); resin acids; unsaturated fatty acids; diterpene alcohols (especially from debarking and mechanical pulping); juvabiones (especially from sulphite and mechanical pulping); lignin degradation products (especially from sulphite pulping); synthetic organics, such as slimicides, oils and greases; and process chemicals, papermaking additives and oxidized metals. The chlorinated organics have been of particular concern, because they are acutely toxic to marine organisms and may bioaccumulate. This group of compounds, including the polychlorinated dibenzo-*p*-dioxins, have been the major impetus for minimizing chlorine use in pulp bleaching.

The amount and sources of suspended solids, oxygen demand and toxic discharges are process-dependent. Due to the solubilization of wood extractives with little or no chemical and resin acid recovery, both sulphite and CTMP pulping generate acutely toxic effluents with high BOD. Kraft mills historically used more chlorine for bleaching, and their effluents were more toxic; however, effluents from kraft mills which have eliminated Cl_2 in bleaching and use secondary treatment typically exhibit little acute toxicity if any, and subacute toxicity has been greatly reduced.

Table: Total suspended solids and BOD associated with the untreated (raw) effluent of various pulping processes

Pulping Process	Total Suspended Solids (kg/tonne)	BOD (kg/tonne)
Groundwood	50–70	10–20
TMP	45–50	25–50
CTMP	50–55	40–95
Kraft, unbleached	20–25	15–30
Kraft, bleached	70–85	20–50
Sulphite, low-yield	30–90	40–125
Sulphite, high-yield	90–95	140–250
De-inking, non-tissue	175–180	10–80
Waste paper	110–115	5–15

Suspended solids have become less of a problem because most mills utilize primary clarification (e.g., gravity sedimentation or dissolved air flotation), which removes 80 to 95% of the settleable solids. Secondary wastewater treatment technologies such as aerated lagoons, activated sludge systems and biological filtration are used for reducing BOD, COD and chlorinated organics in the effluent.

In-plant process modifications to reduce settleable solids, BOD and toxicity include dry debarking and log conveying, improved chip screening to allow uniform cooking, extended delignification during pulping, changes to digestion chemical recovery operations, alternative bleaching technologies, high-efficiency pulp washing, fibre recovery from whitewater and improved spill containment. However, process upsets (particularly if they result in intentional sewering of liquors) and operational changes (particularly the use of unseasoned wood with a higher percentage of extractives) may still cause periodic toxicity breakthroughs.

A relatively recent pollution control strategy to eliminate water pollution entirely is the "closed mill" concept. Such mills are an attractive alternative in locations that lack large water sources to act as process-supply or effluent-receiving streams. Closed systems have been successfully implemented in CTMP and sodium-base sulphite mills. What distinguishes closed mills is that liquid effluent is evaporated and the condensate is treated, filtered, then reused. Other features of closed mills are enclosed screen rooms, counter-current washing in the bleach plant, and salt control systems. Although this approach is effective at minimizing water pollution, it is not yet clear how worker exposures will be affected by concentrating all contaminant streams within the mill. Corrosion is a major issue facing mills using closed systems, and bacteria and endotoxin concentrations are increased in recycled process water.

SOLIDS HANDLING

The composition of solids (sludges) removed from liquid effluent treatment systems varies, depending on their source. Solids from primary treatment principally consist of cellulose fibres. The major component of solids from secondary treatment is microbial cells. If the mill uses chlorinated bleaching agents, both primary and secondary solids may also contain chlorinated organic compounds, an important consideration in determining the extent of treatment required.

Prior to disposal, sludges are thickened in gravity sedimentation units and mechanically dewatered in centrifuges, vacuum filters or belt or screw presses. Sludges from primary treatment are relatively easy to dewater. Secondary sludges contain a large quantity of intracellular water and exist in a matrix of slime; therefore they require the addition of chemical flocculants. Once sufficiently dewatered, sludge is disposed of in land-based applications (e.g., spread on arable or forested land, used as compost or as a soil conditioner) or incinerated. Although incineration is more costly and can contribute to air pollution problems, it may be advantageous because it can destroy or reduce toxic materials (e.g., chlorinated organics) that could create serious environmental problems if they were to leach into the groundwater from land-based applications.

Solid wastes can be generated in other mill operations. Ash from power boilers can be used in road beds, as construction material and as a dust suppressant. Waste from lime kilns can be used to modify soil acidity and improve soil chemistry.

4

Rubber Products

INTRODUCING RUBBER

In 1736, a French astronomer was sent by his government to Peru to measure an arc of the meridian, He brought home samples of the milky fluid and reported that the Indians used it for lighting. He wrote that it burned without a wick very brightly and that the indians made shoes from it which were waterproof. The Indians collected the gummy fluid from trees in pear-shaped bottles on the necks of which they fasten wooden tubes. Pressure on the bottle sends the liquid squirting out of the tube, so they resemble syringes. Their name for the fluid, he added, was cachuchu or caoutchouc.

Thirty-four years later, an English writer wrote about a different use for the tree gum and a new name. A stationer accidentally discovered that it would erase pencil marks, And, as it came from the Indies and rubbed, of course it was renamed India rubber.

RUBBER SOLED SHOES

About the year 1820 American merchantmen, sailing between Brazil and New England, often carried rubber as extra ballast on the home voyage and dumped it on the wharves at Boston. One of the shipmasters exhibited to his friends a pair of native shoes made from rubber. Another, with more foresight, brought home five hundred pairs, and offered them for sale. They were thick, clumsily shaped, and heavy, but they sold. There was a demand for more. In a few years half a million pairs were being imported annually. New England manufacturers bid against one another along the wharves for the tree gum which had been used as ballast and began to make rubber shoes.

Macintosh Elastic Fabric

European vessels had also carried rubber home; and experiments were being made with it in France and Britain. A Frenchman manufactured suspenders by cutting a native bottle into fine threads and running them through a narrow cloth web. And Macintosh, a chemist of Glasgow, inserted rubber treated with naphtha between thin pieces of cloth and evolved the garment that still bears his name.

Waterproof Fabrics

At first the new business in rubber yielded profits. The cost of the raw material was infinitesimal; and there was a demand for the finished articles. In Roxbury, Massachusetts, a firm manufacturing patent leather treated raw rubber with turpentine and lampblack and spread it on cloth, in an effort to produce a waterproof leather. The process appeared to be a complete success, and a large capital was employed to make handsome shoes and clothing out of the new product and in opening shops in the large cities for their sale. Merchants throughout the country placed orders for these goods, which, as it happened, were made and shipped in winter.

Rubber Has Problems in the Heat and Cold

But, when summer came, the huge profits of the manufacturers literally melted away, for the beautiful garments decomposed in the heat; and loads of them, melting and running together, were being returned to the factory. And they filled Roxbury with such noisome odors that they had to be taken out at dead of night and buried deep in the earth.

And not only did these rubber garments melt in the heat. It presently transpired that severe frost stiffened them to the rigidity of granite.

ORIGIN AND HISTORY OF NATURAL RUBBER

Natural rubber from the latex yielding trees *Hevea* and *Castilla* was known to South American Indians in early times (pre-Columbian era). Latex was allowed to evaporate naturally and processed into usable articles (such as balls used in ritualistic games). "Caoutchouc", the world first used to describe the condensed juice of the*Hevea* tree, was the French spelling of an indigenous term for "weeping wood".

Natural rubber was first scientifically described in 1735 by C.-M. de la Condamine, a member of a French geographic expedition to South America. In 1770 the English chemist Joseph Priestley found that it could be used to rub out pencil marks (the term "rubber" was thus coined). Other applications gradually developed for waterproof shoes and clothes.

The world rubber industry began to develop in the 1800s, with the invention of the masticator and the vulcanization process. Demand for rubber grew rapidly with the invention of the solid and later the pneumatic rubber tire and the demand for rubber insulation by the electrical industry. With a view to spreading rubber cultivation, seeds of the *Hevea brasiliensis* were smuggled from Brazil to the United Kingdom in 1876 at the instigation of the British India Office. Seedlings were raised at Kew Gardens and then shipped to Ceylon (Sri Lanka) and Singapore, where H.N. Ridley, director of the Singapore Botanic Gardens, introduced new horticultural and tapping methods. Seeds of the rubber tree were later brought to other tropical regions, especially the Malay area and Java and Sumatra, marking the origin of the rubber plantation industry in Asia.

After Ridley, the next important progress was made in the Netherlands colonial territory of the East Indies (Indonesia), where more sophisticated breeding techniques (selecting high-yielding trees as a source of seedlings for new plantings and budgrafting -that is, vegetatively propagating chosen materials) were introduced.

A notable fact in the early 1900s was also the emergence of production by smallholders, which produced rubber as one amongst several crops. Despite its original commercial development as a plantation crop, by the mid-1930s natural rubber production was evenly split evenly between estates and smallholdings.

The monopoly position of natural rubber as the only kind of industrial elastomer remained virtually unchallenged until World War II (WWII). The origins of synthetic rubber can be traced to the 19th century, when research on the chemistry of natural rubber led to the isolation of isoprene, the chemical compound from which natural rubber is polymerised.

During World War I, under the stimulus of the blockade by the Allies, German scientists produced a crude synthetic rubber, and during the 1920s and '30s several polymerising processes were fisrt developed in Germany, in the Soviet Union, the United Kingdom, and the United States. However, it was not until WWII that synthetics started to be produced on a large scale. Western Europe and the United States were cut off from their main sources of NR, and turned to the domestic chemical industry to increase SR capacity. By the early 1960s synthetic rubbers had overtaken natural rubber in volume.

For more detailed information on the history of NR, please refer to:

- History of natural rubber (the *International Rubber Research and Development Board*);
- "Historical perspectives" at bouncing-ball.com;
- "Natural Rubber", the newsletter of the R-S Information Center for Natural Rubber (articles listed under the heading "history").

CROP PLANT PROFILE

Rubber plantations mainly consist of only one species, *Hevea brasiliens,* a variety of plants of the genus *Hevea (Euphorbiaceae family*), native to Brazil. Commonly known as the rubber tree, *Hevea brasiliensis* is a tall erect tree with a straight trunk and bark which is usually fairly smooth and grey in colour. The plant, growing up to over 40 meters (m) in the wild, characteristically does not exceed 25 m in height when it is under cultivation. Whereas by nature the rubber tree is a perennial (lasting for over 100 years) plant, it is usually replanted after 25-35 years in plantations, when latex yields tend to decrease to an uneconomic level.

Cultivation of *Hevea brasiliensis* in Brazil, its native habitat, was severely hindered by blight in the early 20th century. At present, most of the world's natural rubber is produced by rubber trees descended from Hevea seedlings transplanted from South America to South and Southeast Asia.

The rubber tree flourishes in the tropics with annual rainfall of 2,000-4,000 mm evenly spread throughout the year, and temperatures ranging between 24-28°C. Accordingly, the production of natural rubber (NR) is concentrated in a few tropical countries. However, as a result of intensive breeding programmes, rubber tree areas can be found in locations with an annual rainfall of as little as 1,500 mm per year and a dry season of up to 5 months. The following areas are particularly well suited for *Hevea* rubber culture: the Amazon basin of Latin America; the Thai-Malay peninsula; Sabah and Sarawak in East Malaysia; Sumatra, Java and Kalimantan in Indonesia; Kerala in India and the South-Western part of Sri Lanka; West-Central Africa (*e.g.*, Nigeria, Côte d'Ivoire, Liberia). There are also some areas suitable for NR production in Southern China and Southern Vietnam.

For more detailed information on the rubber tree, please refer to:

- The National Plants Database of the United States Department of Agriculture (USDA, NRCS. 2004. The Plants Database, Version 3.5).

For the world as a whole, the area under rubber plantation has been increasing significantly during 1961-2005 from an estimated level of about 3.9 million hectares (ha) in 1961 to reach over 8 million ha in 2005. However, it should be said that a significant share of the reported rubber area might not be available anymore for latex production.

This potential withdrawal is likely to be attributable to various technical, commercial and political reasons, including over-age (a large portions of the mature areas would need replanting), endemic diseases (for example, a large share of the rubber area in Brazil is thought to be afflicted by SALB) and as a result of political instability (Nigeria, the Democratic Republic of Congo and Liberia).

It should be stressed that Indonesia's share of world total NR planted area is much higher than its share of global production. Indonesia had the largest NR planted area in the world with over 2.67 million hectares in 2005 (comparing with Thailand's 1.68 million hectares). Yet its production in 2005 was 70% of Thai output.

Figures reveal a decline in the NR planted area for many large producing countries. Malaysia reported a steady decline in rubber acreage, which had fallen from the peak of 1.89 million hectares in 1978 to 1.25 millions in 2003 (1.4 in 2004). A similar situation is true for Sri Lanka (the reported figures show a decline in the planted area from 215,100 ha in 1961 to 114,190 in 2005).

In Africa, rubber acreage had declined in Nigeria throughout the decades 1967-1987 (yet the largest area under rubber in Africa in 2005), the Democratic Republic of Congo, and Liberia, mainly as a result of political instability. However, recent changes in Liberia might be a factor leading to the rehabilitation of NR plantation in this country.

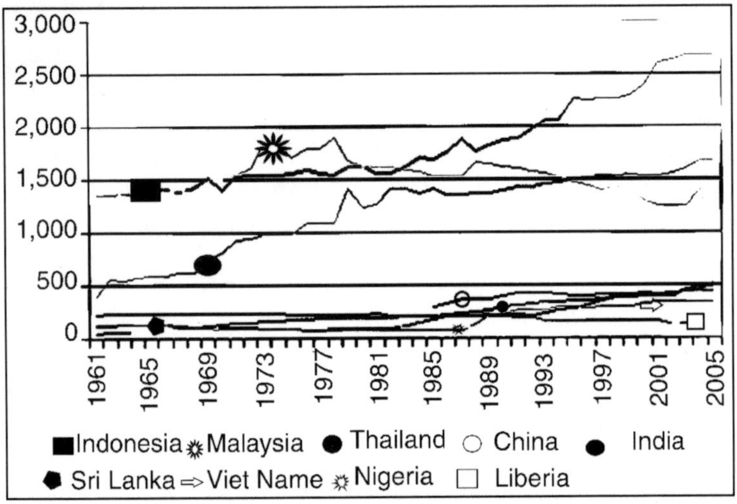

Fig. Development of the World NR Plantation Area (Thousand Hectares), 1961 - 2005

The Declines in NR planted area in many traditional producing countries were more than offset by some increases in planted area in certain large producers (notably, Indonesia, Thailand, China and India). In relative terms many smaller producing countries also performed well. The area in Vietnam has more than doubled from an estimated 85,000 ha in 1981 to around 222,000 ha in 1990. A cooperation agreement with the former USSR resulted in extensive planting of new rubber trees.

The agreement was suspended following the disintegration of the Soviet Union, but the area continued to grow to reach 510,000 in 2005. In Afr ica, Cameroon almost quadrupled the 1960 planted area (from 11,000 ha in 1961 to about 45,900 by 2005); Côte d'Ivoire, now the third largest area under rubber in Africa, recorded the most remarkable expansion (up to 74,000 ha in 2005 from 10,406 in the early 1970s) as a result of large rubber production projects. A few newcomers to the industry can also be seen, such as Gabon and Guinea (where production started in the 1990s).

KINDS OF RUBBER MATERIALS

Rubbers are polymeric materials endowed with the properties of flexibility and extensibility: with the application of force, the molecules straighten out in the direction in which they are being pulled; on release from being extended, they spontaneously recover their normal, random arrangements.

Rubbers include natural rubbers (NRs) and synthetic rubbers (SRs). Natural rubber is a naturally occurring substance obtained from the exudations of certain tropical plants. Synthetic rubber is artificially derived from petrochemical products. Among the most important synthetics are styrene-butadiene, polybutadiene, and isoprene (commonly classified as the "general purpose" SRs), as well as ethylene-propylene rubbers (often referred to as "specialty rubbers"). The prices of these synthetics have been historically in the range of natural rubber prices and their markets have, although to varying degrees, overlapped those of natural rubber. Synthetic rubbers are materials with *ab origine* distinctive chemical structures, whereas the emphasis with natural rubber lies on different types and gradeswithin one single broad category.

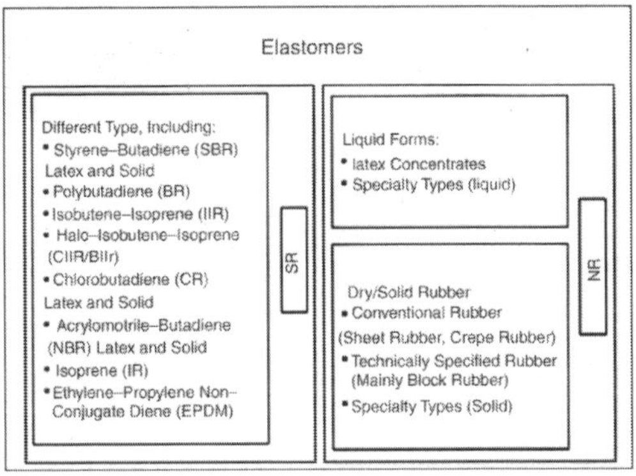

COMPOSITION, TYPES, AND PROPERTIES OF NATURAL RUBBER

Natural rubber (polymer designation cis-1-4 polyisoprene, empirical formula (C5H8n) is obtained from the sap ("latex") of several rubber-yielding plants (*e.g.*, *Hevea Brasiliensis* and *Parthenia argentatum*) by coagulation with chemicals, drying, electrical coagulation, and other processes. Latex is a polydispersed colloidal system of rubber particles in an aqueous phase. With *Hevea* latex, the dry rubber content varies between approximately 28-40%, although it may rise to 45-50% after a long period of non-tapping (notice that, for statistical purposes, figures for natural rubber may include the dry rubber content of latex). The rubber produced from latex contains, besides the hydrocarbon, relatively small quantities of protein, carbohydrates, resin-like substances, mineral salts, and fatty acids.

Table. Composition of Latex

Proteins	2-2.7%	Resins 1.5-3.5%	Sugars 1-2%
Sterol glucosides	0.07-0.47%	Water 55-65%	Ash 0.4-0.7%

As a biological liquid, latex coagulates within a few hours of tapping at the instance of naturally occurring agents ("field coagulum"). Coagulation may be prevented by the addition of preservatives, which offset enzymatic and bacterial influences. The most widely used preservative has been ammonia (the advantage of ammonia lies in the possibility of removing it by aeration for processes where low ammonia content is required).

Natural rubber is either exported as"latex concentrates" or processed into"dry/solid rubber" (in sheet, crepe or block forms). Latex concentrate is made from freshly tapped field latex, uncoagulated; dry rubber is prepared from coagulated field latex (sheet rubbers and pale crepes) or remilled rubber sheets (estate brown, thin brown, thick blanket, and flat bark crepes). Dry rubber is either visually graded (sheets and crepes) or classified by instruments (block rubber). Over the years, technically specified rubber in block form has become increasingly important at the expenses of other rubbers.

TYPES OF NR

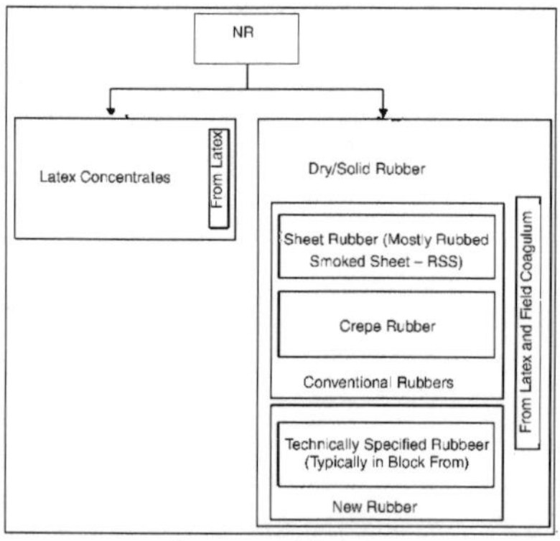

Natural rubber has outstanding resilience and tensile strength, as well as low heat build-up. In addition, natural rubber latex has excellent tack (that is, the ability to stick to itself and to other materials), which makes it best suited for pressure-sensitive adhesives, and excellent water resistance (whereas some synthetics absorb water). However, a drawback is natural rubber moderate environmental resistance to factors such as oxidation and ozone; so too for its scarce resistance to chemicals, including gasoline, kerosene, hydraulic fluids, degreasers, synthetic lubricants, and solvents.

In addition, latex contains proteins that can cause severe allergic response in a small percentage of the population and among medical professionals following extensive exposure. The composition, structure and properties of natural rubber make it an important industrial polymer, suitable for use in many applications.

NATURAL RUBBER PRODUCTION

NR PRODUCTION

NR production varies among geographic regions and within countries, depending also on the type of holdings (estates or smallholdings). On large estates, land is cleared mechanically, whereas on most smallholdings land clearing is still done manually.

Estates and dynamic smallholdings increasingly rely on modern plantation materials and agricultural practices that maximise crop, while traditional smallholders in remote areas often use unselected planting material and follow conventional planting practices. Rubber trees are cultivated at a density of 360-70 trees per hectare (on estates) and 400-500 trees per hectare (on smallholdings).

Depending on conditions, the rubber tree takes 5-10 years to reach maturity -that is, the stage when tapping can be started (the criterion for this is a specific circumference of the trunk, 50 cm at 1 m height). During wintering (lasting for 4-6 weeks), the leaves of the tree die and fall off, and new leaves are formed. The metabolism of the tree and the latex production are substantially affected.

Accordingly, the rubber production fluctuates between months and it is normally low during the rainy season. These seasonal variations are important factors influencing markets. The production cycle of the rubber tree is long and it offers possibilities for joint production with inter-row crops, which would provide income during immature period (popular intercrops in most countries include rice, groundnuts, soya beans, mung beans, pineapple).

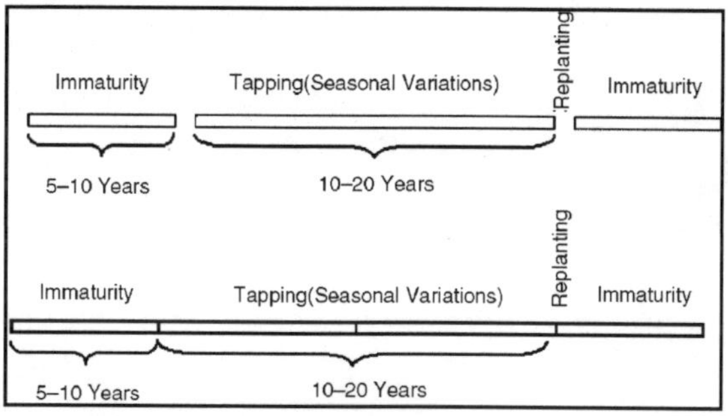

Fig. The Life-cycle of The Rubber Tree

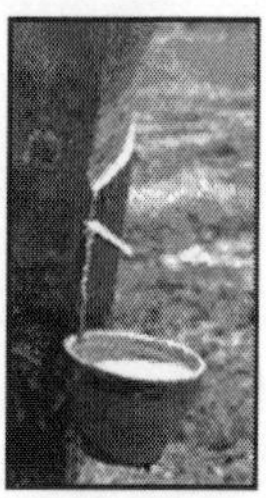

The standard method of tapping involves cutting the bark of the tree with a knife starting at the highest point convenient to the tapper. The cut is generally slant down from right to the bottom left of the trunk at an angle of about 35°. Each subsequent cut is made after the preceding. Generally, rubber trees are tapped about once every two days (each time yielding about 50 g of latex). When the bark of the tree is tapped, thin slivers of bark are excised; the latex exudes from the wound and drips into a cup. With care, tapping may last for 10-20 years and at time even longer (although the intensive tapping of some smallholders is so deep as to severely affect bark renewal capacity and quality). Trees are often rested for a period after heavy tapped.

Latex has a tendency to deteriorate rapidly and coagulate within a few hours of tapping, but it may be preserved with ammonia to limit bacterial contamination ("latex preservation"). A small quantity of latex dries on the bark and is collected the next morning before tapping ("scrap").

Estates *Versus* Smallholders

A major structural change in the NR industry has been the steady decline in estate involvement (as companies switched to less labour-intensive crops such as oil palm) and the increasing importance of the smallholder area. According to some estimates from the International Rubber Study Group (IRSG), smallholdings (by convention, less than 40 ha) accounted for over 78% of world NR production over the period 1998-2003. The balance of planted are and output was accounted for by rubber estates.

Smallholding production edged upward from 5,270 thousands of tonnes in 1998 to 6,310 in 2003 (that is, it grew by an average annual rate of 3%, or about 173 thousands of tonnes). Estate production moved from 1,550 thousands of tonnes in 1998 to 1,670 thousands of tonnes in 2003, or increased by an average annual rate of 1% per cent. It should be noted that the 40 ha threshold is somewhat academic (in most countries the average smallholding size is likely to be 2-3 ha). In addition, statistics on smallhodlings are inevitably less reliable than those available for estates.

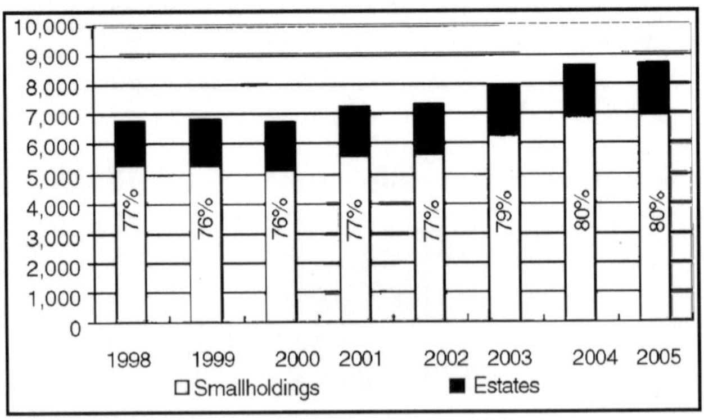

Fig. Output by Estates and Smallholders (Thousand Tonnes), 1998—2005

Whereas most Asian production is smallholder based, estates (relatively large, state-owned plantations) still predominate in the former centrally planified economies (China, Vietnam and Cambodia). The new plantations in Latin America and Africa (where development of the rubber industry started with estates) tend to be on small holdings. NR production is technically and economically viable for both estates and smallholdings. However, there is a technological divergence between smallholder and estate production. Estates capture sizeable economies of scale and vertical integration (as well as some diseconomies -notably, bureaucratic costs, and negative social externalities). They have a serious propension to innovate. By contrast, smallholdings do not enjoy big economies of scale and are in general more conservative. They tend to be less cost sensitive than estates.

Historically, profitability of NR production to smallholders hinged upon a set of technical and structural factors, including the labour-intensive nature of NR production and the availability of low-cost labour, lack of economy in the tapping phase, and supportive social structures. Access to improved planting material is one of the main constraints on smallholdings.

Costs of Producing NR

Unlike synthetic rubber, a capital-intensive industry with few workers, the field production of NR is a labour-intensive sector involving millions of farmers (mostly women). The standard method in tapping is for one person to tap between 200-300 trees in 3-4 hours. The same person then collects the tapped latex. Tapping and latex collection are normally carried out by paid labourers (estates) or household work (smallholdings).

If land values were omitted, labour would constitute the largest single cost per kg in the field production of natural rubber. Although this cost varies considerably among countries (owing to differentials in wages and yields), it tends to be prominent in all contexts. According to data drawn from private records of production and marketing enterprises, management and labour represented over half total direct expenditure on estates and smallholdings in Malaysia in the early 1990s.

YIELDS : SHIFT IN RUBBER PRODUCTION

The upward shift in rubber production coincides with higher productivity, more than expanded area. Productivity increases were achieved by diffusion of tree breeding and selection of technologies (with unselected planting material being replaced by new clonal rubber tree varieties) and by propagating locally suitable trees. New tapping techniques and the use of chemical stimulants also produced major economic advantages, including significant latex outflow and reduced labour costs. For the world as a whole, yields increased from less than 5,500 hg/ha in 1961 to about 11,000 hg/ha in 2005. Indonesia farmers still largely rely on unimproved rubber seedlings and conventional rubber-growing methods (which also entail shifting cultivation with subsistence food crop). Even on estates the level of skills (including lack of technological awareness) has been for long inferior to that in the other major producers and has contributed to relatively lower output per hectare. Effective replanting policies and economic incentives fundamentally explain the better performance of Thailand, where government intervention in the rubber sector has been extensive. In India, government incentives and good educational levels have stimulated widespread replanting of improved varieties, which has steadily increased yields. There has also been large investment in supportive infrastructures, including research.

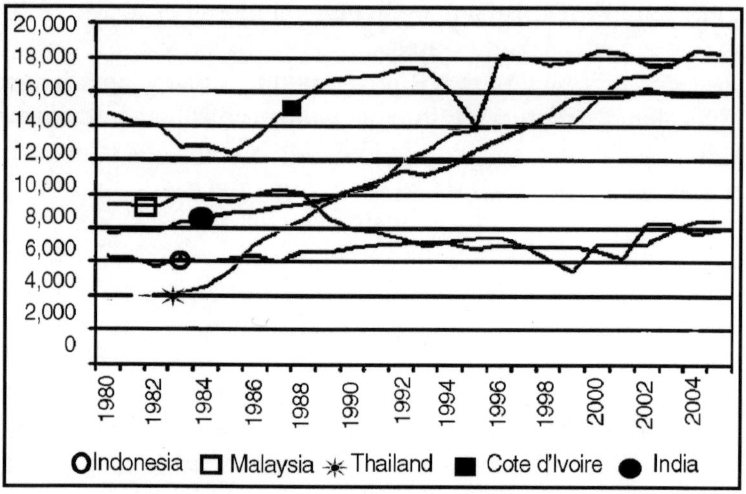

Fig. Nr Yields In Selected Countries (Hg/Ha), 1980-2005

New higher yielding varieties and other technological improvements have offset, to some extents, the effects of increases in labour costs and in the cost of purchased inputs. Nevertheless, as a result of low prices and returns to growers, some producers abandoned tapping.

PESTS AND DISEASES

The most severe disease is South American Leaf Blight (SALB). This disease is endemic throughout the rubber growing areas in the Americas. It also poses a major global threat. The fungus can be controlled by a number of fungicides. Some species are not susceptible to the disease, and some strains of susceptible species are resistant. Crown budding or grafting of resistant plants onto productive trees can be used to control spread.

Corynespora cassiicola is in many ways similar to SALB: it is also caused by fungal attack and it is capable of causing very serious damage. Unlike SALB, it is prevalent in much of Africa and Asia (the disease is especially severe in Indonesia and Sri Lanka), as well as the Americas. Management practices include replanting with resistant trees but some resistant varieties have become prone to attacks. The tapping panel dryness (TPD) syndrome of rubber is due to physiological disorders in the bark tissue (drying up of the rubber producing tissues which can be induced by overexploitation of the rubber tree). It causes the reduction or ultimately total cessation of latex flow upon tapping.

QUALITY

The distinction between industry, national, and international standards is for sake of conceptual clarity only. Industry standards are incorporated in national and international (ISO) standards. Testing for compliance under most domestic schemes shall follow ISO methodologies. ISO standards are used as such or implemented through incorporation in the national standards of individual countries.

Differences in types and grades are of major commercial significance. Indeed, the NR market is highly fragmented by types and, within each types, grades, with sizeable variations in trade flows and price movements.

Industry Standards

Two main grading systems exist:

1. Natural rubber graded by the conventional visual standards given in the "Green Book" of the International Rubber Quality and Packing Conference" (Visually Graded Types)
2. Natural rubber graded by technical specifications (Technically Specified Rubbers)

Specialty grades have been separately developed for rubbers with specialty applications.

Visually Graded Types

Conventional types of rubber (that is, rubber processed into sheet rubber and various crepes) are still visually inspected and graded according to quality and packing specifications set in the "International Standards of Quality and Packing for Natural Rubber Grades (The Green Book)".

These standards were first issued and then revised within the framework of the "International Rubber and Packing Conference", comprising the world's leading rubber manufacturers, growers and trade associations. The grading system was based on the classification first developed in 1928 by the US "Rubber Manufacturers Association" (then "Rubbers Association of America").

Sheet and crepe rubbers intended for export shall be classified into one of the following grades:

Table. Standard International Grades of NR

Form	Types	Grades
Sheet rubber	Ribbed Smoked Sheet (RSS)	1X, 1–5
Crepe Rubber	Pale Latex Crepe (PLC), thin and tick	1X, 1–3
	Sole Crepe (SC)	–
	Estate Brown Crepe (EBC), thin and tick	1X–3X
	Thin Brown Crepe	1–4
	Thick Blanket Crepe	2–4
	Flat Bark Crepe	Standard, hard
	Smoked Blanket Crepe	Pure

Ribbed Smoked Sheets (RSS)

Ribbed smoked sheets are graded on a visual basis as RSS IX, RSS 1, RSS 2, RSS 3, RSS 4 and RSS 5. After grading the sheets are packed in 50 kg bales, with the grades marked on the bales.

For more detailed information on RSS grading, please refer to the International Grade Descriptions for Sheet Rubber (Rubber Board, India).

Crepe Rubber

Crepe rubber is either made from latex ("Latex Crepes") or processed from cup lumps, scrap and cuttings of ribbed sheet rubber ("Field Coagulum Crepes", or "Remilled Crepes"). Latex crepe is graded as "Pale Latex Crepe" (PLC, from fresh field latex) and "Sole Crepe" (SC, obtained from dried PLC). There are 8 grades under the PLC category, with the purest being 1X (the difference between tick and thin grades is only in thickness). Pale and sole crepes are more specialised than remilled crepes, and yield a higher premium. With both latex crepes, and especially the sole crepe, the colour represents an important element in marketing and pricing (a higher premium being obtained with a lighter coloured product).

Remilled crepe rubber prepared from field coagulum materials fall into five categories (further graded on a visual basis):

1. Estate Brown Crepe (EBC), thin and tick (made from lump and other high grade rubber scrap (field coagulam) generated on rubber estates)—6 grades
2. Thin Brown Crepe (Remills), (derived from wet slab coagulum, unsmoked sheets and cup lumps)—4 different grades
3. Thick Blanket Crepe (Ambers), (wet slabs, unsmoked sheets or other high grade scraps)—3 grades
4. Flat Bark Crepe, (processed from all types of low grades scrap including earth scrap)—2 grades
5. Pure Smoked Blanket Crepe, (exclusively derived from ribbed smoked sheet or cuttings)

Technically Specified Rubber (TSR)

The conventional grading system outlined above had many disadvantages, in that it was based on subjective assessments and only recognised visual distinctions that could not have much technical significance. A major step towards evolving a more sophisticated classification for rubber was the development of a grading system strictly based on technical specifications. The move away from visual grading to technical specifications originated in Malaysia in the 1960s with the "Standard Malaysian Rubber" (SMR) scheme. This prompted producers in other countries (notably, Indonesia and Thailand) to develop similar technical specifications for grades. "Technically Specified

Rubbers" (TSR)—that is, rubbers graded on the basis of technical parameters—now account for the bulk of traded rubber. Standardised international contracts for such rubber were first adopted by the International Rubber Association at its inaugural meeting on 24 September 1971.

Rubber is technically specified according to the following sets of rules, which are incorporated and detailed in the national standards of individual countries:

- A– Technical specifications: TSR is graded on technical properties, rather than on visual distinctions. Technical parameters for grading are expressed as either maximum or minimum thresholds (detailed under national TSR schemes). For each grade, the dirt content, ash content, volatile matter content, nitrogen content, plasticity, and colour are tested by instruments. Testing for compliance shall follow ISO test methods.

ISO has specified 6 different grades of TSR, namely, TSR L (high quality and light coloured rubber prepared from latex), TSR CV (viscosity-stabilised high quality latex rubber), TSR 5 (good quality latex rubber, darker than TSR L), TSR 10 and 20 (good quality grades derived from field coagulum, suitable for general purpose uses), TSR 50 (up to 0.50% wt dirt content). The specifications and characteristics of TSR are detailed in the following table.

- B– Packing specifications (specifications with regard to nominal dimensions and weight of bales and crates, as well as technical requirements for wrapping materials).
- C– Registration (suppliers of TSR are required to apply for registration with the competent authority; TSR for shipments is to be accompanied by technical specifications issued by authorised laboratories; bales of TSR must be marked with the registered TSR symbol and the authorised grade).

The prime objective of the scheme was to standardise the marketing of new forms of natural rubber (block rubbers). However, conventional types of rubber (sheet rubber and crepe rubber) may also be graded on technical properties if they are supplied according to packing specifications and meet technical requirements.

Latex Concentrate

Unlike sheet and crepe rubber, latex concentrate is graded according to detailed technical specifications (as for block rubber). Technical specifications are separately established by both the American Society for Testing Materials (ASTM) and the British Standards Institution (BSI). National and international standards are substantially identical in their requirements.

Commercial latex is nowadays mainly prepared by centrifugation, with only two types made on a large scale:

1. HA (high-ammonia) latex, preserved with 0.7% ammonia

2. LA-TZ (low ammonia) latex, preserved with 0.2% ammonia + 0.025% TMTD/ZnO +0.04—0.05% lauric acid as ammonium laurate.

NATIONAL STANDARDS

Individual rubber producing countries are in charge of setting the acceptable limits for each grade of rubber they produce. The main rubber producing countries have their own schemes patterned after the TSR scheme. National schemes conform to specifications laid down by the International Standards Organisation (ISO).

Table. Comparison Chart

TSR CV	Constant viscosity grade	TSR L	Latex rubber
TSR 5	Sheet rubber	TSR 10	Field grade material
TSR 20	Field grade material		

- Malaysian Rubber (Technical note, Department of Statistics Malaysia);Department of Standards Malaysia (subject index: standards for "Rubber", "Rubber latices", "Rubber latex concentrate")
- Indonesian Rubber (The R-S Information Center for Natural Rubber)
- Rubber Grades, India (M/s J&J Trade Links corporate page); Indian Rubber(The Rubber Board, statutory body constituted by the Government of India for the overall development of the rubber industry in the country)
- Papua Rubber Regulation

ISO STANDARDS

To date, ISO's work has resulted in some 127 international standards for "rubber"; 55 standards for "latex and raw rubber"; 30 standards for "rubber compounding ingredients". Standards have been enacted for all the following fields:

ISO International Standards are voluntary, unless incorporated into domestic law. In practice, because they are developed in response to market demand, ISO standards tend to have wide industry application irrespective of whether they have been incorporated into domestic law or not.

SOURCES OF COMMERCIAL NATURAL RUBBER

Commercially, natural rubber is obtained almost exclusively from *Hevea brasiliensis*(genus *Hevea*, family *Euphorbiaceae*), a tall softwood tree indigenous to Brazil. Industry botanists have concentrated their efforts mostly on this species.

Other trees, shrubs, and herbaceous plants produce natural rubber. Among the species which have received commercial attention are the *Ficus elastica* (*Moraceae*family), the guayule bush (of the family Asteraceae), and the Russian dandelion (*Asteraceae*, alt. *Compositae*). Guayule (*Parthenium argentatum*), a rubber-containing desert shrub of the family Asteraceae, native to the north-central plateau of Mexico and the Big Bend area of Texas, is the only other plant under cultivation as a commercial rubber source.

In recent years, these and other species have attracted increasing attention, because of diminishing acreage of *Hevea* plantations, increasing demand for high quality rubber with specialty applications, and severe allergic response to the latex based products of the *Hevea*. Guayule has attracted increasing interest as a commercial alternative for hypoallergenic latex. In a number of Latin American countries, hybridisation between *Hevea benthamiana* and *Hevea brasiliensis* has provided relatively SALB-free sources of rubber. The molecular mechanism of rubber biosynthesis in other plants (for example, in the fig tree -Ficus carica) is being investigated with a view to developing alternative rubber crops.

Hevea brasiliensis and most of the other rubber-yielding species grow only within a well defined area of the tropical and sub-tropical countries, though there are also some rubber-bearing species whose distributional range includes temperate areas (*e.g.*, *Taraxacum kok-saghyz*, native to the Soviet Middle Asia, *i.e.* Kazakhstan and Kyrgyzstan, and China, as well as other species of the family *Compositae*).

For statistical purpose, rubbery materials extracted from species other than *Hevea Brasiliensis,* including guayule (*Parthenium argentatum*) and gutta-percha (*Palachium gutta*), are conventionally referred to as "Natural Gums".

MARKETING CHAINS OF NATURAL RUBBER

Production of natural rubber (NR) is regionally concentrated in South and South East Asia; yet the bulk of consumption occurs in North America, Europe, Japan, and China. The marketing network for NR has thus the crucial role of linking the regional concentration of producers on the one hand and consumers on the other. It entails a multitude of dealers, selling and buying agents, brokers and exporters, who perform discrete but related functions at various stages of the chain.The overall marketing chain may be examined by considering the internal and external marketing separately.

INTERNAL MARKETING: ESTATES V. SMALLHOLDINGS

Rubber production is carried on in both estates and smallholdings. These two NR producing sectors possess discrete organisational structures and marketing channels.

Marketing of Estate Rubber

NR production and processing is fully integrated in most of the bigger estates. Moreover, a few large estates have moved downstream into rubber goods manufacturing and other activities. Some major players in the rubber plantation industry are now diversified industrial conglomerates which derive a small share of their income from production of the raw product (consider, for example, the Guthrie group in Malaysia). Marketing structures vary greatly between and within different geographical locations.

A few large estates have direct links to consumers and, in such cases, direct sales are carried out. There may be either direct consignment to selling agents in consuming countries (as done by agency houses) or sales to consumers' buying agents who operate locally.

The balance of estate NR is handled through open trading. Essentially in this structure, estates rely on dealers and brokers operating both locally (primary market) and in consuming countries (terminal markets). Estates usually retain ownership over the product until the final stage. Moreover, large estates normally sell forward.

Marketing Channels for Smallholders' Rubber

The arrangements in the marketing of smallholders' NR depend largely on the size of the holding and its integration into group schemes.

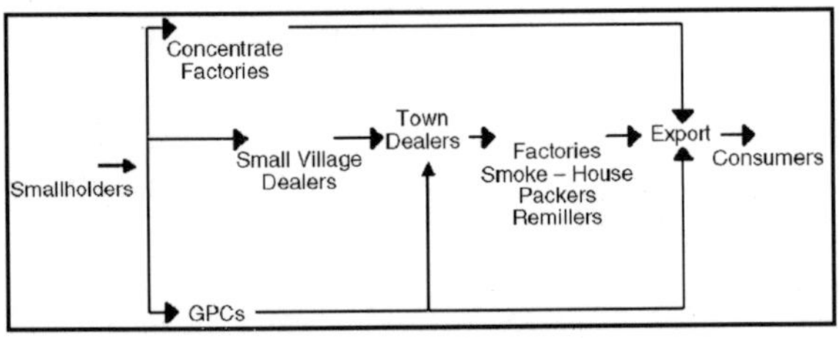

Fig. Marketing of Smallholders Rubber

Traditionally, the flow of NR from producers to consumers has involved a chain of intermediaries operating at different levels: local rubber dealers (village level), middle dealers (town level), remillers, packers, and exporters (international). Under this system, the village dealer, often a shopkeeper and a moneylender in addition to being a rubber dealer, often represents the smallholder's only link in the marketing chain. The point has been made that this set-up may lead to monopsonistic and monopolistic situations.

First level traders and their agents collect rubber in sheet (most often unsmoked sheets) and crepe forms from smallholders, sort and sometimes smoke it, and sell it to "middle" dealers. From the middle dealers the rubber is delivered to wholesale dealers and exporters, who sort, grade and pack it before shipment against f.o.b. or c.i.f. contracts.

Strategies for the development of the smallholder sector included processing via group processing centres (GPCs). The GPC's members are able to benefit from an increase in scale and sophistication in rubber processing. They tend to produce better quality sheets, which are then sold to the middle- and upper-level network of traders (bulk sales through tenders).

External Marketing: Open and Direct Trading

Rubber commodity markets provide a channel of supply from producers to final consumers via dealers and brokers (open trading). In addition, a few large producers may have direct links to consumers and, in such cases, direct sales are carried out (direct trading).

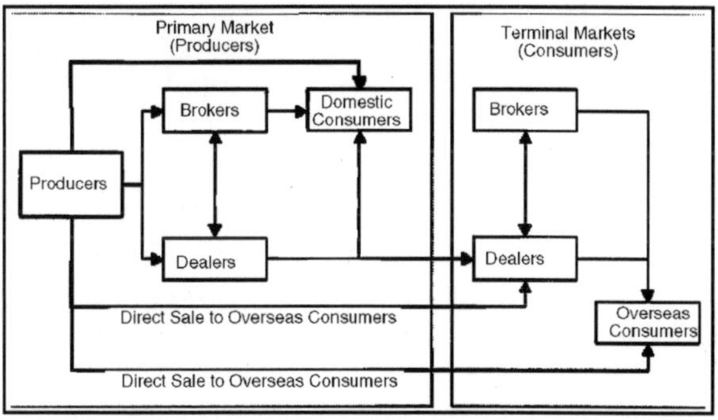

Fig. Marketing Channels (NB: Bad Printed Quality)

Open Trading

In the earlier structure of NR trading, dealers and brokers openly traded rubber through"primary markets" in Kuala Lumpur and Singapore. Rubber from these primary markets was then forwarded to counterpart"terminal markets" in London, New York, Shanghai, Amsterdam, Hamburg, Paris, and Tokyo. Primary and terminal markets perform an important function in matching supply and demand and discovering prices. Two types of trading strategies occur in these markets: spot or cash transactions involving trade in physical commodities and futures trading.

The primary markets in Kuala Lumpur and Singapore are regulated and administered by the Malaysian Rubber Board (MRB) and the Rubber Association of Singapore (RAS), respectively. Singapore is primarily an"entrepôt" market (storage and re-packaging): it imports rubber for re-export in consuming countries. Primary markets provide services related to the transfer of ownership and the physical flow of rubber from producers to the port of shipment.

Terminal markets located in Hamburg, Shanghai, Amsterdam, Paris, and Tokyo are basically import markets supplying the local needs. Similarly, New York is largely a domestic market to supply the needs of the United States (to a lesser extent it imports for re-export to Canada and Latin America). By contrast, the London market is primarily an"entrepôt" market, in addition to supplying domestic requirements, national rubber can also be re-exported.

NR handled through open trading notably involves less specialist grades (*e.g.*, TSR 20, RSS1 and RSS3) and crops from less developed regions, particularly in Africa. The few dealers transacting most of the rubber on open markets are actually subsidiaries of large companies. Major trading firms in the London market included Symington and Son. Ltd. (along with Symington Italia Gomma Lattice of Rome, part of the Guthrie group which controls large plantations in Malaysia) and Cargill plc (a subsidiary of the giant grain marketing company, with trading offices in Singapore and Tokyo).

Direct Trading

Direct trading between producers and consumers (especially tyre-making consumers) has been encouraged by further developments in the rubber industry (notably, requirements for better consistency and tailoring to consumer needs and the growth of rubber goods manufacture in East Asia).

Direct trading may occur via sales to consumers' buying agents operating in producing countries (for example, the rubber-buying offices set up in Singapore by most of the international tyre companies). Yet there are also other direct outlets. In particular, a few large estate companies and smallholder group organisations have deepened and extended their marketing links with particular consumers, sometimes operating through selling agents in consuming countries.

NR direct trading between big suppliers and consumers typically entails long-term contracts of up to 18 months. Direct trading arrangements often involve specialty rubbers tailored to consumer needs.

Marketing Structures in Malaysia, Indonesia, and Thailand

Malaysia

The two NR producing sectors, *i.e.* estates and smallholdings, are largely mingled in location (unlike Indonesia, where estates and smallholdings tend to have distinct locations). Throughout history, Malaysia has exhibited relatively higher labour and land costs in NR production than Indonesia and Thailand. The organisational structure of NR production in Malaysia well reflects the above conceptual outline (twofold producing sector -estates v. smallholders; external marketing by way of direct sales and open trading). Following progressive concentration, the estate sector is dominated by a few large companies supplemented by smaller estate business (a few medium-sized foreign firms and a network of smaller domestic units, often performing service activities). The largest estate companies are incorporated as a public limited company and listed on the Kuala Lumpur and Singapore stock exchanges. The Malaysian government maintain substantial interests in major plantation companies previously owned by foreign (chiefly British) interests.

Estate companies have developed into major corporate conglomerate with substantial interests in various sectors of the Malaysian economy. A few of them entail overseas operations and distribution centres in consuming countries. Most companies have developed close links with international concerns, which still prominently feature in downstream rubber goods manufacturing (foreign investors include Goodyear in tire manufacturing and Safeskin Corporation for the manufacturing of latex concentrates into dipped goods).

The organisational structure of the Guthrie group well illustrates the extent to which Malaysian estate companies have moved downstream and diversified horizontally. Only a small share of income is currently derived from rubber plantations.

In the smallholding sector, government programmes have involved establishing group smallholdings through the public agency Federal Land Development Authority (FELDA). In addition to implementing land development programme throughout the country, FELDA is itself a major estate company with downstream involvement in rubber processing and trading (through a hub of wholly and majority owned subsidiaries). The Malaysian Rubber Development Corporation (MARDEC) traditionally processed natural rubber from Malaysian smallholders into various commodities. The figure below details the marketing structures for Malaysian smallholders rubber.

Today, Mardec is a major diversified corporate group with downstream linkages and a trading arm. In addition to factories that produce various categories and grades of natural rubber, Mardec subsidiaries custom-make a wide range of value-added rubber and plastic products that meet customers' specifications. R1 International, the Singapore-based Mardec's marketing and trading arm, deals with local and foreign traders. Direct trading links are also established by accredited Mardec agents in consumer countries all round the world. Mardec is the local partner in joint-venture companies such as with Ansell of Australia (rubber gloves) Fillatice of Italy (latex thread) and Cabot of Australia (carbon black).

Indonesia

Sumatra is the chief natural rubber area. Significant further cultivation districts are located in western and central Kalimantan. Unlike Malaysia, smallholders are largely confined to remote areas, especially in the key NR growing area of northern Sumatra. There is a persistence of traditional production technologies, as reflected in relatively low yields. The marketing of smallholders' rubber is still largely carried out through a chain of dealers who often enjoy monopsonistic positions vis-à-vis individual smallholders.

Efforts have nonetheless been made to group together smallholdings. For example, under the Nucleus Estate Scheme smallholders were provided with small plots of high yielding tree crops. Participating farmers in a determined location shared the benefits of centralised technological and managerial assistance. Indonesian estates include large publicly owned plantations and private estates. As in Malaysia, most estates also grow oil palm and other crops. Although increasing involvement in downstream rubber processing, there is still relatively little rubber goods manufacturing (dominated by Bridgestone and Goodyear). Publicly owned plantations (the "PTP" plantations) were based on Dutch concerns that were nationalised in the late 1950s and early 1960s. Initially structured as public entities, they were subsequently incorporated under private law as limited liability companies, although publicly owned. Some estate companies still combine private and public attributes. In 1996 the three PTP companies were consolidated into the State-owned enterprise "PT.

Perkebunan Nusantara III" (Persero). Persero is a vertically integrated concern with core activities in palm oil, rubber, and cocoa production and processing. It owns 4 rubber processing facilities (rubber thread factory, dipping process factory, rubber article factory, and cyclized rubber resin factory) and entertains direct sale linkages with foreign consumers, notably through overseas marketing offices. Private estates comprise both Indonesian companies and foreign concerns.

TECHNOLOGY OF NATURAL RUBBER PROCESSING INTO DIFFERENT CATEGORIES

Freshly tapped latex and field coagulum can be made into various rubber categories by different processing methods. Traditionally, NR has been made into sheets whose quality was visually inspected. In addition to sheets, NR has also been traded ascrepes (similarly graded on a visual basis). The traditional methods of processing into sheet and crepe rubbers are still widely used, especially by smallholdings. Block rubber (graded by technical tests) was first introduced by Malaysia in the mid-1960s.

Sheets

Ribbed smoked sheets (RSS) production from latex occurs in either factory-scale facilities (on estates and larger smallholdings) or in tiny field units (individual smallholdings). Despite its variability in scale, it always includes the following operations: blending, coagulation, milling, drying, and finishing. The layout for RSS production (factory-scale processing) is detailed in the figure below.

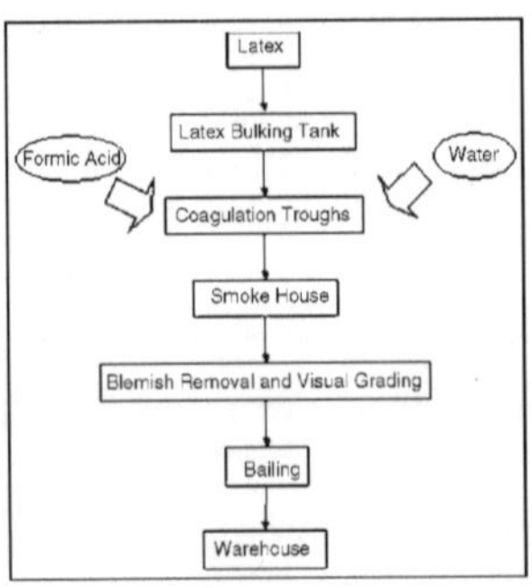

Fig. Ribbed Smoke Sheet Plant

Latex from various sources is initially collected and blended in a large latex bulding and blending tank. This blending operation is important to ensure the production of uniform and consistent rubber. Coagulation is effected by the addition of coagulants, such as formic or acetic acid. In factory-scale processing it occurs in coagulation troughs, where latex is first diluted with water. Individual smallholdings blend and coagulate their latex in 4-5 litre pans. Milling involves feeding slabs of coagulum to successive pairs of rollers ("sheeting battery") to produce sheets of uniform thickness.

Appropriately grooved rollers imprint on each sheet "ribs", which expand the surface area for drying. After sizing, the sheets are hung on lines and dried in smokehouse at successively higher temperature ("ribbed smoked sheets"). On individual smallholdings, sheets are often dried by natural ventilation ("white unsmoked sheets") and sold to a dealer who smokes and grades them. Blemishes are then removed manually and sheets are visually graded. They are finally pressed into bales with talc to prevent adhesion. The process of making sheet rubber is relatively simple and it is still commonly used on smallholdings and smaller estates.

Crepes

Rubber in crepe form is processed from either latex (pale and sole crepe) or cup lumps, scrap and poor quality sheet rubber (remilled crepes). The traditional method of rubber processing to produce crepe rubber is similar to that for sheet rubber. An important additional step in making crepe rubber is the removal of yellow carotenoid pigments in the latex. In addition, the latex is coagulated by "fractional" coagulation: the first fraction is the unstable, yellow fraction processed into off-colour pale crepe of relatively low grade; the remaining latex is a wither material. The coagulum formed is then washed and fed to pairs of rollers rotating at different speeds, which manufacture the rubber into thin crepes. The crepes are dried in hot drying chambers or tunnels or on drying floors.

Block Rubber

Since the 1960s, new processes have been developed to produce technically graded rubber from latex and field coagulum. The production of block rubbers involves relatively sophisticated machineries and heavy power consumption. The basic layout of block rubber production from latex is detailed in the figure below (factory-scale processing).

In block rubber production from latex, latex arriving from different locations is first blended in a large bulking tank. Chemicals are added to control viscosity and affect colour. The latex containing the requisite chemicals is then coagulated by adding coagulants (formic acid) in a long coagulation trough. Solid latex coagulum is then processed into crumb by either physical or chemical means.

In the physical case, the coagulum is first fed though the rotating rolls of a creping machine.

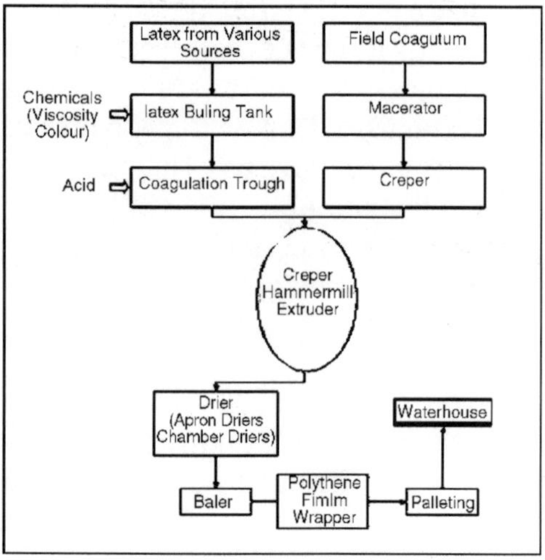

Fig. Block Rubber Plant

The crepe is then mashed into small pieces through a hammer mill and finally concerted into crumb by an extruder. Alternatively, incompatible oil is added to the latex in the coagulation trough. The material is then fed though a creper and transformed into crumb. In both cases, the crumbs finally obtained are dried by hot air. Depending on the throughput, factories may use single-layer tray driers with direct firing burners, apron driers, or chamber driers. The dry rubber is finally baled (commonly by means of a hydraulic press) and wrapped in polythene to prevent adhesion between bales in the crate. When field coagulum rather than latex is being processed into block rubber, the materials involved include foreign matter and require pre-cleaning. Field coagulum materials are first forced through a macerator (to slice up large blocks) and then passed through a series of creeping rolls. There is obviously no need for coagulation.

Preserved Latex Concentrates

Latex concentrates are processed by various methods, including centrifugation, evaporation, creaming, and electro-decantation.

RG MANUFACTURING

Important steps in the manufacturing of rubber goods from latex and dry rubber include:

- *Preparing* (NR generally requires prior mastication and other preparatory operations before mixing with compound ingredients)
- *Compounding* (the mixing in the rubber inputs of materials additional

to the base polymer or polymer blend and vulcanization constituents, to meet increasingly diverse requirements in rubber goods)

- *Shaping* (for example, the rubber mix may be heated and shaped into a mould of appropriate form, or frictioned on to a fabric)
- *Vulcanisation:* Vulcanization was one of the key discoveries in the manufacture of rubber products. It involves a chemical process whereby individual polymer molecules are linked to other polymer molecules by atomic bridges (the actual chemical cross-linking is usually done with sulfur, but there are other technologies, including peroxide-based systems). Vulcanization makes the raw rubber harder, more durable and more resistant to chemical attack.

ECONOMIC POLICIES

Market regulation has been extensive in the natural rubber (NR) sector. It included the international buffer stock schemes envisaged under the 1979, 1987 and 1995 lateInternational Natural Rubber Agreements (INRA). After the demise of the INRA, it is still embodied in collective attempts at managing production by the International Tripartite Rubber Organisation (ITRO).

INTERNATIONAL NATURAL RUBBER AGREEMENTS

The first International Natural Rubber Agreement was signed in 1979 under the auspices of the United Nations Conference on Trade and Development (UNCTAD) and then renegotiated in the mid-1980s (1987 Agreement) and during 1994-1995 (1995 Agreement). The 1979 Agreement had as members seven exporting countries accounting for about 95% of world exports and 25 importing countries plus the European Community (three-fourths of world imports at the time).

The objective of the 1979 agreement was essentially to reduce excessive price fluctuations around the trend in rubber market prices. Exporting countries also stressed the need to support prices at levels considered remunerative to producers and to help to stabilise export earnings and increase these earnings based on expanded export volumes. Basic differences between importing and exporting countries in their conception of the role of a commodity agreement for NR led to wide divergences in negotiations, particularly as regards price levels. Both the 1987 and the 1995 Agreements maintained the same objective as in the 1979 Agreement.

The 1979 Agreement provided for a buffer stock of a maximum size of 550,000 tonnes (400,000 tonnes of normal buffer stock and 150,000 tonnes of contingency buffer stock) as the sole instrument of market intervention for price stabilisation (the Agreement therefore excluded export quota or production controls). The scheme entailed the purchase and selling of the buffer stock when actual NR quotations (the daily market intervention price (DMIP), as defined in the Agreement) were outside upper and lower price bands in relation to a reference price. The reference price was initially set at 210 Malaysian/Singapore cents per kilo. The DMIP was a composite average of daily official current-month prices on the Kuala Lumpur, London, New York and Singapore markets (RSS1, RSS3 and TSR20 grades of rubber; all quotations converted to f.o.b.

Malaysian/Singapore ports and expressed in a composite Malaysian/ Singapore currency). Intervention price levels (where the buffer stock manager had discretion to act) were set at +/- 15% of the reference price level. The Agreement also established upper and lower trigger action prices at +/- 20% of the reference price, which the buffer stock manager was obliged to defend.

The buffer stock was financed by direct cash contributions by governments (although the contingency buffer stock could be financed by cash borrowed form commercial sources).

The costs of the buffer stock were shared by importing and exporting countries (within each group, contributions were assessed in proportion of each members' share in NR trade).

The Agreement was managed by the International Natural Rubber organisation (INRO), headquartered in Kuala Lumpur, Malaysia. The main operational provisions in the 1987 and 1995 Agreements were much the same as those of the first Agreement.

There were, however, several modifications to the provisions for the periodicity of reviews and size of revision of price levels triggering market interventions.

The reference price on entry into force of the 1987 Agreement was set at 210.66 Malaysian/Singapore cents per kilo (206.68 at the entry into force of the 1995 Agreement). In 1998 the exporting Member Countries of INRO asked for an increase of the reference price with 5% in relation to the economic/ currency crisis in South-East Asia. The importing Member Countries rejected this proposal. Following the withdrawal of Malaysia, Thailand and Sri Lanka, in September 1999 the Council of INRO decided to terminate the third International Natural Rubber Agreement (INRO was liquidated in December of the same year). It was the last "survivor" of all stabilization commodity agreements.

It is generally recognised that the buffer stock operations under the 1979 and 1987 Agreements succeeded in their price stabilisation objective. As regards the achievement of prices at levels considered remunerative to producers, there was widespread criticism in the exporting countries of the low level of the reference price (set at 210 Malaysian/Singapore cents per kilo when the first Agreement entered into force in 1980; down at 206.68 fifteen years later).

For a review of the main features of the Agreements, please refer to this report (pdf, 1.10 MB) by the UNCTAD secretariat.

LATEX ALLERGY

Latex, also known as rubber or natural latex, is derived from the milky sap of the rubber tree, found in Africa and Southeast Asia.

Latex allergy is an allergic reaction to substances in natural latex. Rubber gloves are the main source of allergic reactions, although latex is also used in other products such as condoms and some medical devices.

WHAT IS LATEX

Natural rubber latex comes from a liquid in tropical rubber trees.

This liquid is processed to make many of the following rubber products used at home and at work:

- Balloons
- Dishwashing gloves
- Waistbands on clothing
- Rubber toys
- Pacifiers and baby-bottle nipples
- Rubber bands
- Adhesive tape and bandages
- Diapers and sanitary pads
- Condoms

In addition, many medical and dental supplies contain latex, including gloves, blood pressure cuffs, urinary catheters, dental dams and material used to fill root canals, as well as tourniquets and equipment for resuscitation. Non-latex substitutes can be found for all of these latex-containing items.

The protein in rubber can cause an allergic reaction in some people. This reaction can range from sneezing to anaphylactic shock, which is a serious condition that requires immediate medical attention. The thin, stretchy latex rubber in gloves, condoms and balloons is high in this protein. It causes more allergic reactions than products made of hard latex rubber (such as tires). Because some latex gloves are coated with cornstarch powder, latex protein particles can stick to the cornstarch and fly into the air when the gloves are taken off. In places where gloves are being put on and removed frequently, the air may contain many latex particles.

WHAT ARE THE SYMPTOMS OF LATEX ALLERGY

Latex allergy can be mild or severe, with symptoms such as:

- Itchy, red, watery eyes
- Sneezing or runny nose
- Coughing
- Rash or hives
- Chest tightness
- Shortness of breath

Some people who wear latex gloves get bumps, sores, cracks or red, raised areas on their hands. These symptoms usually appear 12 to 36 hours after contact with latex. Changing to non-latex gloves, using glove liners, and paying more attention to hand care can help relieve these symptoms. A person who is highly allergic to latex can also have a life-threatening allergic reaction, called anaphylactic shock.

Symptoms include:

- Difficulty breathing
- Dizziness
- Confusion
- Wheezing
- Nausea
- Vomiting
- Rapid or weak pulse
- Loss of consciousness

Someone having an anaphylactic reaction needs immediate medical attention.

LATEX ALLERGY

Allergy to latex was first recognized in the late 1970s. Since then, it has become a major health concern as an increased number of people in the workplace are affected.

Health care workers exposed to latex gloves or medical products containing latex are especially at risk. It is estimated that 8-12% of health care workers are latex sensitive. Between 1988-1992, the Federal Drug Administration (FDA) received more than 1,000 reports of adverse health effects from exposure to latex, including 15 deaths due to such exposure.

As used in this topic, latex refers to the natural rubber latex manufactured from a milky fluid that is primarily obtained from the rubber tree (Hevea brasiliensis). Some synthetic rubber materials may be referred to as "latex" but do not contain the natural rubber proteins responsible for latex allergy symptoms.

Latex allergy is addressed in specific standards for the general industry.

Latex products are everywhere. Anyone can develop a latex allergy. Health care workers and rubber industry workers seem to have the highest risk for latex allergy.

Health care workers who have hay fever have an especially high chance of developing a latex allergy, as 25 per cent of all health care workers who have hay fever show signs of being sensitive to latex. People also at risk are those who have had many operations (especially in childhood), people who have spina bifida (a birth defect that affects the development of the spine) and people who have a food allergy.

LATEX ALLERGY AND FOODS

Because some rubber proteins are similar to food proteins, some foods may also cause an allergic reaction in people who are allergic to latex. The most common of these foods are banana, avocado, chestnut, kiwi fruit and passion fruit. Although many other foods can cause an allergic reaction, avoiding all of them might cause nutrition problems. Therefore, it's recommended that you avoid only the foods that have already given you an allergic reaction.

See a doctor, preferably one with experience in treating latex allergy. Your doctor will take a detailed history and may confirm the diagnosis with a blood test. Skin testing is usually not used to test for latex allergy, except in some specialized centers. It can cause severe reactions if it isn't done by an experienced person.

Although there is no treatment for latex allergy, you can reduce your risk of reaction by avoiding direct contact with latex. Take steps to find out which products in your environment contain latex. Then, find substitutes you can use for those products. It's also important to avoid breathing in latex particles from powdered gloves or other sources. If you are a health care worker or a patient, everyone around you should wear powder-free latex gloves or non-latex gloves. If you are a health care worker, compare different kinds of non-latex gloves to find the ones that are best for you. Always wear or carry a medical alert bracelet, necklace or keychain that warns emergency medical technicians (EMTs) and doctors that you are allergic to latex. Talk to your doctor about getting a prescription for an epinephrine self-injection pen, to use in case of a serious reaction. You may wish to carry non-latex gloves with you all the time for use by emergency personnel if you need medical attention. If you are exposed to latex at your job, tell your employer and co-workers about your latex allergy. Avoid latex gloves completely if you're not at risk for blood and body fluid contamination. Use powder-free gloves if latex gloves are preferable. These measures will help keep others from becoming allergic to latex.

HOW CAN I LEARN MORE ABOUT LATEX ALLERGY?

Take steps to educate yourself and others by joining the resource networks and support groups listed below. Work to support workplace policies, industry practices and government legislation that will support the safe use of latex and non-latex alternatives.

Over 90% of natural rubber comes from the latex, or milky sap, of the rubber tree—Hevea brasiliensis. Latex is used in a large number of products including adhesives, foam, carpet backing and in a number of "dipped" products. Dipped products include surgical gloves, household gloves, catheters, condoms, balloons, etc. Natural latex rubber is also found in toys, erasers, sports equipment, clothing, elastic bands, and many medical and dental devices.

For reasons not yet clearly understood, some individuals can become allergic to certain constituents of the latex molecule. Once allergic, you may remain sensitive for the rest of your life. Growth in the latex allergy problem has been related to the rising demand and use for latex products. The onset of recommendations for universal precautions resulted in increased use of latex gloves—to meet the increased demand, manufacturing processes may have changed resulting in a more highly allergic product.

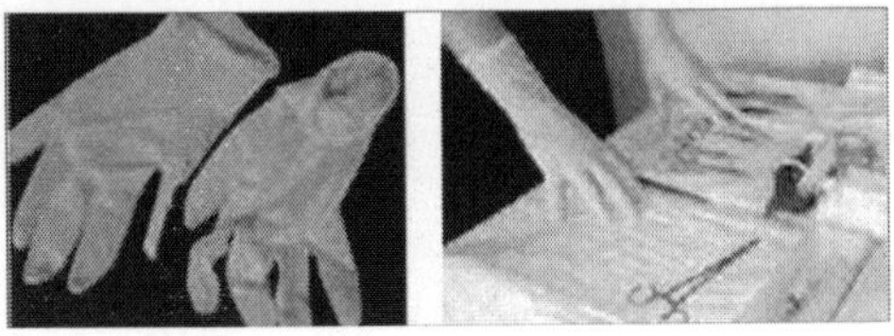

WHO IS AT RISK

Anyone who is exposed to latex may become sensitised—fortunately, the majority of individuals do not.

However, certain populations are more at risk:

- Rubber-industry workers
- Health care workers (from wearing rubber gloves)
- Individuals with congenital urinary anomalies (from use of catheters and other latex devices) *e.g.* children with spina bifida.
- Individuals with a history of sensitivity to certain foods *i.e.* banana, avocado, chestnut, tomato, peach or kiwifruit
- Individuals who have undergone multiple medical/surgical procedures
- Individuals with a history of non-medication related anaphylactic reactions during anaesthesia

WHAT CAUSES LATEX ALLERGY?

The exact cause of latex allergies is unknown. It appears that repeated and frequent exposure to latex and rubber products may bring on symptoms in some people.

Since the late 1980´s there has been a dramatic rise worldwide in allergy to latex. A logical explanation is the use of universal precautions for preventing the spread of infectious diseases such as the AIDS virus. As a result, the use of latex gloves is nowadays widespread. Frequent exposures to latex and rubber products are common. Health care workers are at particular risk for latex allergy.

According to the AAAAI, over 10% of healthcare workers are thought to have a latex allergy, as well as more than half of all spina bifida patients. Between 1% to 6% of the general population are allergic, experts believe.

Allergic people's immune systems identify latex as a pathogen - a substance or organism which harms health. Experts say that susceptible people react to a protein in the sap of the rubber tree. The immune system triggers cells in the body to produce IgE (immunoglobulin E); these are antibodies which fight the latex component. The next time the body comes into contact with latex, the IgE antibodies sense it and signal the immune system to release chemicals, including histamine into the bloodstream.

The more a susceptible person is exposed to latex, the greater their immune system is likely to be - this is called sensitization.

TYPES OF REACTIONS TO LATEX

The common reactions of latex sensitive individuals exposed to latex are contact urticaria, dermatitis and asthma. Contact urticaria usually presents with itching and swelling of the skin at the site of contact with latex. This may, for example, be hands from wearing gloves, genitals from contact with condoms, etc. The symptoms usually start within 5—15 minutes after coming into contact with the latex article, although it can be delayed for several hours. Symptoms can continue for a variable period, from several hours to days after the latex contact has ceased. Contact dermatitis from latex may take several days to appear. It presents with an itchy, scaly rash, although there may be small blisters if the reaction is acute. The rash will usually last several days to weeks but if exposure to latex continues, the rash will last longer. Contact dermatitis is not generally caused by sensitivity to latex protein but rather to the chemicals used in the manufacture of the latex product, including antioxidants and rubber accelerators *e.g.* thiuram, carbamates, MBT, etc. Immediate-type hypersensitivity requires previous sensitisation and is the most potentially dangerous reaction to latex. Clinical presentations vary but may include contact urticaria, coryza, conjunctivitis, stinging or burning, asthma, and, with mucosal or parenteral exposure, anaphylaxis. A common cause for the asthmatic reaction is powdered gloves. The starch powder picks up the latex proteins and when the gloves are removed the powders can then be inhaled or come into contact with the skin on the face where it can cause an allergic reaction.

EXPOSURE METHODS

Contact through skin, mucous membranes, parenteral, intravenous and inhalation.

How to Diagnose Latex Allergy

In most cases, a diagnosis of latex allergy can be made from an accurate history and clinical examination. There are a number of skin tests which can confirm latex sensitivity, including patch tests (for allergic contact dermatitis type reactions), skin prick test (for contact urticaria type reactions) and radioallergosorbent assay blood test (RAST) which is not as accurate as the skin tests.

GUIDELINES FOR HEALTH CARE WORKERS

Help prevent latex sensitivity:

- Use appropriate work practices to reduce chance of reactions to latex. When wearing latex gloves, use only approved barrier cream. After removing latex gloves, wash hands thoroughly and dry thoroughly.
- Practice good housekeeping: frequently clean areas and equipment contaminated with latex-containing dust.
- *Learn to Recognise the Symptoms of Latex Allergy:* Skin rash; hives; flushing; itching; nasal, eye or sinus symptoms; asthma; and (rarely) shock.
- Care should be exercised when removing latex gloves to reduce the risk of exposure of latex allergens to self and/or colleagues.

LATEX ALLERGY IN THE HEALTH CARE WORKER

- Avoid contact with latex gloves and products. Commonly used latex medical products include: rubber gloves; elastic bandages; adhesive tape; urinary catheters; electrode pads; wound drains; stomach and intestinal tubes; urinary collection devices; protective sheets; enema tubing and tips; dental cofferdams; rubber pads; and fluid circulating warming blankets. Equipment containing latex includes: haemodialysis and anaesthesia equipment; rubber masks; electrode pads; head straps; rubber tourniquets; rubber nasal-pharyngeal airways; rubber oral-pharyngeal airways; teeth protectors; bite blocks; blood pressure cuffs; rubber breathing circuits; reservoir breathing bags; rubber ventilator hoses and bellows; rubber endotracheal tubes; latex cuffs on plastic tracheal tubes; latex injection ports on intravenous tubing; and certain epidural catheter injection adaptors.
- Avoid areas where you might inhale the powder from latex gloves worn by other workers.
- Tell your employer and health providers (physicians, nurses, dentists, etc) that you have latex allergy.
- Wear a medical alert bracelet.
- Consult with your doctor to determine whether you should carry an emergency adrenaline (epinephrine) kit in case of accidental exposure to latex and/or in the event of anaphylactic shock (adrenaline immediately counteracts the shock).

PATIENTS WITH LATEX SENSITIVITY

The following guidelines are summarised FDA advice to health professionals to ensure a patient's latex allergy is not overlooked:

- Include questions about latex sensitivity when taking a patient's health history—asking patients whether they have ever experienced itching, rash or wheezing after wearing latex gloves or inflating a toy balloon may be useful. Spina bifida patients are at higher risk for acquiring latex allergy, as are others who have undergone numerous medical procedures. As well, patients with sensitivity to certain foods (chestnuts, kiwifruit, bananas, avocado, peaches, figs, passion fruit) should be considered high risk for latex allergy.

- Those patients who report signs of latex allergy should have the allergy recorded in the clinical notes.
- Ensure patients are informed when they have had a suspected latex-related allergic reaction while under the professional's care, and recommend a latex allergy test to those individuals.
- *A latex-safe environment should be offered to patients with latex allergy:*
  - No-one in the area should wear powdered natural rubber gloves
  - Only non-latex gloves should be used to directly examine the patient
  - Do not inject or withdraw fluid through rubber ports or intravenous lines
  - All natural rubber latex products that routinely directly or indirectly come into contact with mucosa, non-intact skin, and internal bodily spaces (*e.g.* catheters, endotracheal tubes) should be removed or made latex -safe (*e.g.* covering—blood pressure cuffs used over clothing, stethoscope tubing covered with stockinette)
  - Products made of dry-moulded rubber (*e.g.* wheelchair tyres, tool handles) do not need to be removed from the area but if direct contact with such objects is routine, precautions should be taken to minimise direct exposure by covering the object or body part exposed (*i.e.* having the patient wear nonlatex gloves while wheeling self in wheelchair)
  - Hospital dietary services should have latex-safe meals that lack food known to cross-react with antibodies to natural rubber latex (*e.g.* avocado, banana, potato, tomato, papaya, chestnut or kiwifruit)
  - If a latex-safe operating room is not always available, patients should be scheduled as the first procedure of the day so that latex-allergens can not build up in the environment
- For patients with an identified latex allergy *e.g.* alert in clinical notes, medical alert bracelet, etc. staff will need to have latex-free products available. It is recommended that each ward/department consider which of their patients are likely to be "high risk" and develop an emergency kit of latex-free products for use (acknowledging that there may be other patients not identified as "high risk" who may develop

a latex allergy). The equipment required will vary and will need to be identified by the clinical area.

THE MAIN USES OF NATURAL RUBBER

Because of its elasticity, resilience, and toughness, natural rubber (NR) is the basic constituent of many products used in the transportation, industrial, consumer, hygienic and medical sectors. Of these major end-use markets for rubber, transportation is by far the largest single sector, with tyres and tyre products accounting alone for over 50% of NR consumption. Truck and bus tires would represent the largest single outlet for NR, followed by automobile tires. General rubber goods for commercial and industrial use account for the balance. These non-tyre rubber items include industrial products (for example, transmission and elevator belts, hoses and tubes, industrial lining, and bridge bearings); consumer products (like golf or football balls and other recreational and sports goods, erasers, footwear and other apparel); and articles for use in the medical and health sector (notably, condoms, catheters and surgical gloves) as well as seismic materials (for instance, over 500 and 2,500 buildings are respectively fitted with seismic rubber bearings in China and Japan). Latex articles (typically condoms, gloves, threads, adhesives, and moulded foams) could be included in different categories in terms of end-use.

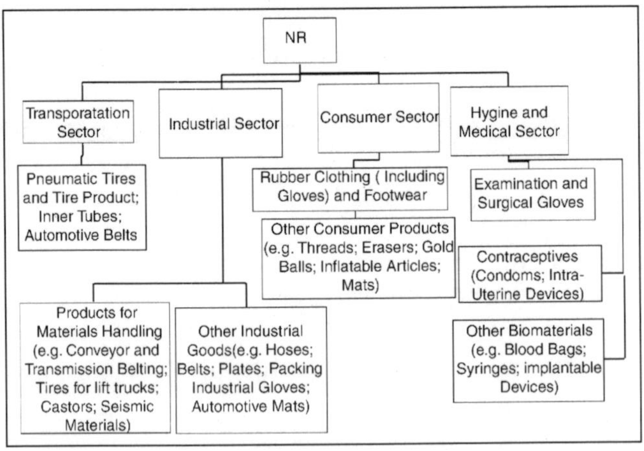

Fig. NR Major end Uses

Growth in demand for certain end products conventionally using rubber does not necessarily translates into growth of NR consumption, owing to competition from synthetic rubber (SR) and new materials.

NR AND SR IN TIRES

Despite the competition of synthetic compounds, natural rubber continues to hold an important place in tire consumption. In particular, its superior tear strength and excellent resistance to heat up makes it better suited for high-performance tires used on racing cars, trucks and buses, and aircraft. In these applications, the potential for switching from natural to synthetic rubber is quite limited, given the clear-cut technological advantages to natural rubber.

Rubber tires are of two types: solid (or cushion) tires, in which the rubber portion functions to carry the load and absorb shocks; and pneumatic tires, with compressed air that fills the tire. The former are used on industrial machinery and on military vehicles; pneumatic tires are used for almost all free-moving vehicles (*i.e.*, other than railroad cars). Pneumatic tires include tires for automobiles, trucks and buses, (motor) bicycles and airplanes and "off-the-road" tyres for special vehicles (such as construction vehicles and agricultural machinery). The distribution channel for pneumatic and solid tyres is two-tired, consisting of original equipment manufacturers (OEMs) and the replacement market.

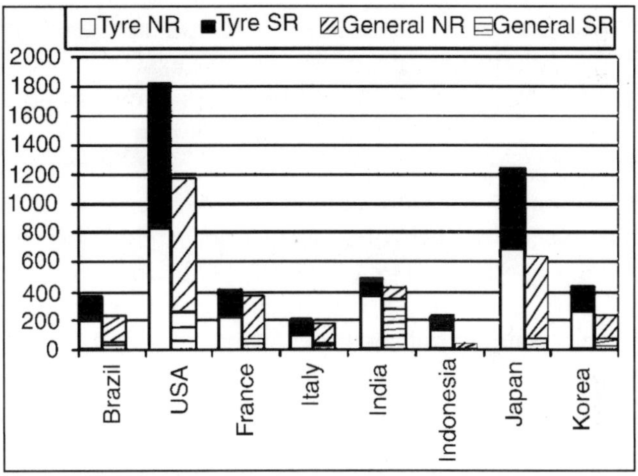

Fig. Breakdown by Major Sectors (Nr and Sr)

Separate rubber compounds are used for different parts of the tire. Generally, the larger the tire the greater the share of natural rubber. The table refers to relatively small-size truck tires.

Table. Typical Tire Composition by Weight

Passenger tire	**Average weight: New 25 lbs, Scrap 20 lbs.**	
Natural rubber	14%	
Synthetic rubber	27%	
Carbon black	28%	
Steel	14–15%	
Fabric, fillers, accelerators, antiozonants, etc.	16–17%	
Truck Tire	**Average weight: New 120 lbs., Scrap 100**	**lbs.**
Natural rubber	27%	
Synthetic rubber	14%	
Carbon black	28%	
Steel	14–15%	
Fabric, fillers, accelerators, antiozonants, etc.	16–17%	

TECHNICAL FACTORS OF LATEX PRODUCTS

As for high-performance tires, technical factors (performance needs of the products and process technology) also constrain the ease of substituting SR for NR in the latex product market. For example, because it is waterproof (whereas some synthetics absorb water), NR latex is best suited for use in surgical and medical examination gloves and in condoms. NR latex is possibly the best protection against pathogens such as HIV.

Latex products include, *inter alia*, condoms, gloves, threads, adhesives, and moulded foams. They found applications, including specialty applications, in different sectors, among which is the medical and hygiene sector.

Overall, the competitive relation between natural and synthetic rubbers hinges upon a variety of related factors, which include technical factors (technical properties of the rubber, performance needs of the product, process technology), economic factors (relative input prices and processing costs), and relative efficiency of marketing channels. As in the case of contraceptives and high-performance tires, the extent to which substitution from NR to SR may take place on the basis of price is limited. It is mostly dependent on technical factors, which also determine the category and grade of elastomers used.

NR is processed in different types and grade with characteristic applications. For example, latex concentrate is the basic constituent of contraceptives, surgical dipped goods, and rubber threads. Ribbed smoked sheets are used in the industrial sector when extra tough rubber is needed (for example, for tank liners). Pale Crepe is valuable for medical sundries, footwear, cements and adhesives.

Rubber producers increasingly custom-make a wide range of special rubbers that meet customers' specifications. These rubbers are often made by means of compounding natural rubber and synthetics, which also tends to complicate the distinction between NR and SR.

NATURAL RUBBER

Natural rubber is processed latex, which is a naturally-occurring compound found in 10% of all flowering plant species. Natural latex is most commonly harvested from the Para Rubber Tree, which produces latex in quantities sufficient for industrial harvesting, though most of the rubber now used in the world is synthetic. Natural latex farms feature many rows of tapped rubber trees from which latex slowly seeps out.

There are few uses for untreated natural latex; in order to become rubber tires, gloves, adhesives, catheters, diaphragms, water seals, engineering components, hoses, industrial bands or any of the other widely used natural rubber products, latex must be treated and shaped. There are several treating and curing processes that can be applied to raw latex, the most common of which is vulcanization.

Vulcanization is the processes by which latex is heated and combined with additives like sulfur and carbon black to reduce its stickiness and increase its elasticity and strength properties. Once shaped, vulcanized or subject to other treatments, rubber products can be used in applications in industrial, commercial and consumer products contexts. Aerospace, automotive manufacturing, construction, healthcare, food processing, sporting goods, defence contracting and many other operations make extensive use of natural rubber products.

The word latex is actually the word for any stable dispersion of any polymer microparticles in a liquid; the natural latexes that can be extracted from plants are only one kind of latex.

The words latex and rubber are often used interchangeably, though this is incorrect. Natural Para Rubber Tree latex becomes rubber only after processing. The latex is extracted from the tree and combined with a stabilizing agent like ammonia to prevent the coagulation (clumping) to which latex is given when exposed to atmosphere.

The ammonia-latex mixture is then strained and concentrated in a processing facility; its water content is decreased using a centrifuge. At this point, the latex-stabilizer mixture is called latex concentrate. It can then be heated, at which point it becomes solidified to a degree that depends on its water and stabilizer content and the extent to which it is heated. This semi-solid product is now the most basic form of natural rubber, though in this form it has very few uses. Rubber extruders, injection molders and other rubber processing equipment are necessary to shape the rubber into usable products, and additional treatment processes like vulcanization are usually necessary after shaping to bring the rubber's qualities of strength and durability to acceptable levels.

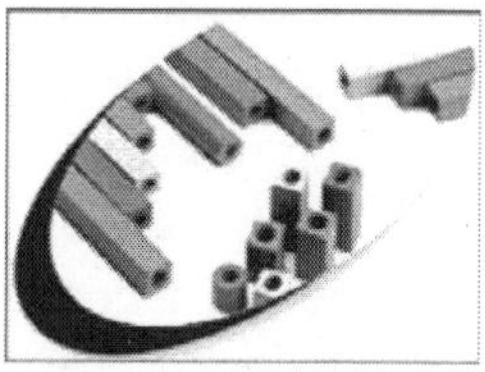

Fig. Natural Rubber Image Provided by Timco Rubber Products, Inc

MARKET OF PRODUCTION IN RUBBER

The supply of natural rubber (NR) hinges upon the interplay among several factors, including production capacity, underlying technological change, input and processing costs as well as price differential with synthetic rubber (SR). During 1961-2005, NR production grew on average 3.4% a year. By the early 1960s synthetic rubbers had overtaken natural rubber in volume. In 2005, the six leading producers -Thailand, Indonesia, Malaysia, India, China and Vietnam—accounted for roughly 89% of world NR production. Combined output in Thailand, Indonesia and Malaysia alone represented around 70% of the global output. Despite the substitution effect with SR, world natural rubber production increased from roughly 2,1 million tonnes in 1961 to over 9,1 million tonnes in 2005. According to IRSG's predictions for 2009, the world's production rise for NR should not overcome 3.6% and 3.5% for SR. As more, Indonesia should stay the most dynamic market with a rise of 6.8%.

Production in Thailand has been increasing steadily during 1961-2005, from a reported 186,100 tonnes in 1961 to over 3 million tonnes in 2005 (at an average rate of 6.5% per year over the period). It has tended to stabilize for the last two years due to bad weather conditions (heavy rains). Thailand nevertheless remains the world leading NR producing country. Its share of world output has also increased over the years, from as low as 8.8% in 1961 to over 33% in 2005.

Indonesia has become the second largest producer behind Thailand. NR production in Indonesia has been rising during 1961-2005 to reach 2.128 million tonnes by 2005 (+4.8%), at an average rate of 2.6% per year. It should be noticed however that its share of world output has declined slightly over the years, from as high as 32.7% in 1961 to 23.3% in 2005. This was caused by much sharper rise in the output of the largest producing country, Thailand, as well as increases in India and China.

Malaysia had been the largest NR producing country for most of the 20th century, until it was relegated to third position in the early 1990s. Back in the mid-1970s, Malaysia accounted for almost half of the world's NR output. Following the steady decline in production since the second half of the 1980s, its share dropped to 11.5% in 1999. It has been argued that as the Malaysian economy advanced, the cost of opportunity of producing NR increased. Higher wages in the city attracted workers, which made labour scarce as well as land and capital relatively expensive for NR production.

Consumption

Rubbers are intermediate goods used in producing final consumer goods (notably tires). In this respect, demand for a specific type and grade of rubber is derived, since it depends on demand for determined final goods, of which rubbers are only one component. General income, expected price of a specific type and grade of NR and of its substitutes, expected price of final goods, as well as relative processing costs for NR and SR can all be regarded as having an impact on NR consumption.

The five largest NR consuming countries are China, the United States of America (USA), Japan, India, and Malaysia. World rubber's consumption was of approximately 21 millions of tons in 2005 (+3,1%). The growth is less significant compare to the last four years (especially 2003 with a peak of 6%). IRSG estimates that the world rubber's consumption should rise of 4% per year until 2009 to reach 24.5 million of tons. China overtook the USA to become the world's top NR consuming country. According to figures from the International Rubber Study Group, NR consumption in China grew quite sharply by an average 12% a year throughout 1998-2005 to reach over 1.8 million tonnes in 2005. High demand for motor vehicles, and hence for tires, stimulated by the stronger economic growth, together with the spike of oil prices which resulted in higher price for synthetic rubber, contributed to the relatively sharp rise in NR consumption in China.

The USA relinquished its top position in NR consumption since 2001, with continued sharp growth in Chinese rubber consumption and a decline in US consumption (over the 1998-2001 period, -6% per year according to IRSG). As for the other largest consumers, most rubber goes to the tire sector. In the USA, synthetic rubber has traditionally been the major type of rubber consumed.

In Japan, NR consumption allegedly grew during 1998-2005 at the relatively modest rate of 2.8% a year. The decline in the ranking of Japan's rubber consumption was not only a result of the rapid growth of China's rubber industry, but also because of Japanese rubber manufacturers relocating elsewhere.

Synthetic rubbers (SR) overtook NR in production and consumption volume since the 1960s. Growth in natural rubber (NR) usage has been stronger than that of synthetic rubber (SR) for 1998-2005 (a growth rate of 4% -annual average- as compared to 3% for SR). NR consumption indeed grew of 5.5% in 2005 compare to 1.3% for SR consumption, which led to a ratio of 57.7% (the weakest since 1964).

Sharply rising oil prices, which may add significantly to the costs of major SR producers, and developments in NR pricing relative to SR (the relative NR/SR% price ratio rose from around 80% to 120% between the start and finish of 2005) have intensified the debate surrounding possibilities for substitution.

According to ISRG's, the growth in rubber's consumption predicted until 2009 should be more absorbed by NR (4.7%) than SR (3.6%). The ratio for SR would therefore decrease again to reach 57.2% in 2009.

In China (the largest single consumer of NR), SR consumption increased at an estimated 14% a year over 1998-2005 to reach 2.6 million tonnes by 2005 (up from 1 million tonnes in 1998); NR consumption has been increasing since 1998 at 12% per year, *i.e.* from 839 thousand tonnes in 1998 up to around 1.8 million tonnes in 2005.

Trade Flows

NR Exports

Thailand is the largest exporter. Following the trend of production, Thai NR exports increased at about 7% per year during 1961-2004 to reach around 3 million tonnes in 2004 (from as low as 184,500 tonnes in 1961). In 2004, NR exports from Thailand accounted for 40% of the world total as compared to 8% in 1961.

Indonesian NR exports also paralleled trends in production (exports have increased on average by 2% per year during 1960-2004). In the 1990s Indonesia's share of world exports reached its highest peak during the Asian financial crisis in 1998 at 31% (the reason behind the relatively sharp rise in Indonesian exports during this period was identified in the relatively weak currency). Over 1999-2004, Indonesia accounted for an average 26% of world exports. Malaysian NR exports have declined in both absolute and relative terms.

In 2003, NR exports dropped to about 946 thousand tonnes or 14% of the world total in 2003, compared with 1.66 million tonnes and more than 50% in 1977. This sharp decline in exports follows the trend in production. It should be stressed also that an increasing share of the declined rubber output has been consumed domestically. Malaysia reported a stronger export performance in 2004 (52% increase over the previous year), which may suggests a partial reversal of the downward trend. It is interesting to note Vietnam emerging as an exporter of some note.

In terms of NR exports by type (dry NR versus NR latex), in Indonesia, Malaysia and Thailand the share of dry rubber to all shipments was 99%, 92% and 81%, respectively (average during 1999-2004).

For all major exporters, available trade data show an increase in technically specified rubber (TSR) at the expense of sheet grades (RSS) and latex. According to some estimates from the IRSG (RIR, March 2004), in Indonesia the share of TSR exports to all shipments increased significantly (from 41% to 96%) during the period 1973-2003, while the shares of RSS and latex declined to 3% and less than 1%, respectively. Malaysia also behaved in a similar way (according to IRSG, the shares for 2003 would be 91% for TSR, 8% for latex and 1% for RSS). In Thailand, RSS exports relative to all shipments declined from 84% to 47% during the period 1973-2003; over the same period, the shares of TSR, latex and others would have risen to 34%, 15% and 4%, respectively. The share of RSS in Thailand remains remarkably high compared to Malaysia and Indonesia. This may be partly due to the fact that most NR in Thailand is produced by smallholders.

NR Imports

Mainland China emerged in 2003 as the world's largest natural rubber importer and kept its leading position with NR imports reaching more that than 1.4 million tonnes in 2004, nearly 32% more than in 2002. This increase was fuelled by ongoing growth in demand for automobiles coupled with stagnation in domestic natural rubber production. In Malaysia there has been a rise in imports as a result of the rapid growth in domestic demand for NR combined with a decline in domestic NR production.

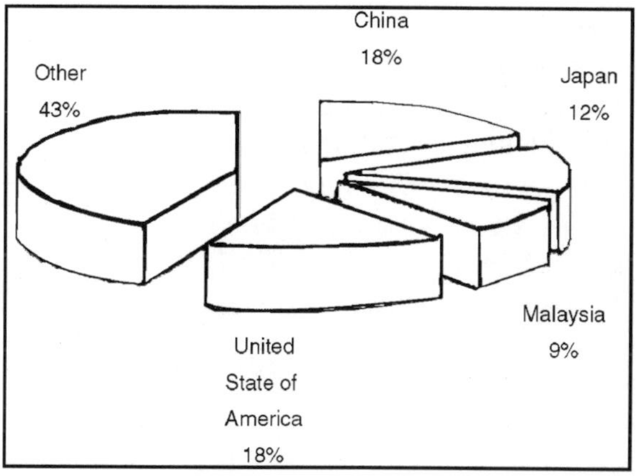

Fig. Breakdown of Imports by Major Importing Countries (%), Average 1999-2004

Most NR trade is still interregional (from South-East Asia to North America and Europe), although the increasing consumption of South-East Asian NR in Asia is enhancing the share of that region.

INTERNATIONAL TRIPARTITE RUBBER ORGANIZATION

In 2001, the three most important producers, namely Indonesia, Malaysia and Thailand, established the International Tripartite Rubber Organization (ITRO) in order to manage rubber production in an effort to maintain orderly market growth and guarantee a minimum price to their domestic producers. Subsenquantly, with the entry of Vietnam, ITRO's members account for more than three-quarters of world exports of natural rubber.

To support NR prices ITRO's members agreed to reduce production (of 4%) and exports (of 10%). The reduction in output would reduce global stocks, which should,*ceteris paribus*, have a positive effect on rubber prices. At the same time they established an organizational structure, the International Tripartite Rubber Organization (ITRO), to collectively manage their production for the next few years. To stabilize world natural rubber prices, they launched in October 2003, the International Rubber Company Ltd (IRCO), sometimes know as "the rubber consortium", to pool resources of the three charter members. Vietnam was invited to joint this company and its membership could be further expanded (other countries that are on the invitation list are India, Sri Lanka, Papua new Guinea as well as Singapore). IRCO management would step in to buy rubber when prices declined.

SOCIO-ECONOMIC, ENVIRONMENTAL AND POLICY ASPECTS

A large number of rubber smallholders are subsistence farmers working in small family groups. They tend to face many disadvantages and problems, which minimise their chance of achieving good economic returns from their holdings.

Key issues include the following:

- They have little, if any, access to modern plantation materials and do not follow agricultural practices that maximise crop. Furthermore, they often lack sufficient funds to fertilize.
- They are dependent on family labour, which limits the scale and productivity of the farming undertaken.
- They tend to have poorer land and uncertain land tenure (early smallholders emulated activities on nearby estates and planted rubber trees on cleared forest land to which they did not have title). In subsistence agriculture, a formal title to the land may be rare, especially where the law refers to customs, and communities decide in practice on individual land occupancy (also, entitlement to land may be restricted by legislation).

Government programmes have evolved establishing group smallholdings and implementing land development programmes. In addition, various price- and income-support mechanisms have been conceived at the domestic level. For instance, the objective of the Indian Price Stabilization Fund (PSF) is to provide income support to rubber growers when price of rubber falls below remunerative level. A price band of 40% has been adopted with a price spectrum band of +/- 20% from the reference price (the seven years moving average of international prices). When the price falls below 20% of the reference price, growers can avail themselves of the relief. If the price exceeds the 20% threshold, growers have to contribute.

The Scheme is based on the principle of contributions from the growers and from the Government depending on normal/boom/distress periods. Each participating grower has to open an account with an initial deposit of ₹100 and a non refundable contribution of ₹500. When price remains within the band, the grower will be required to deposit ₹500 each year to his account; the Trust Fund also will contribute ₹500 as matching contribution to his account. When price falls below the lower bound of the band, the Trust Fund will deposit ₹1,000 to his account and the grower need not make any remittance.

The implementation of the scheme will be monitored by a high-powered committee leaded by Secretary/Additional Secretary (Plantations), Department of Commerce (Ministry of Commerce and Industry, Government of India). Large quantities of water are required during processing of rubber (mainly for washing, churning and dilution).

The factory effluent contains rubber particles and other substances which may give rise to serious oxygen depletion, if discharged into public water courses without preliminary treatment. In many rubber-producing countries, rubber factories are required to comply with the discharge standards laid down by the competent authority.

5

Cultivation and Harvesting of Tea

Camellia sinensis is an evergreen plant that grows mainly in tropical and subtropical climates. Some varieties can also tolerate marine climates and are cultivated as far north as Cornwall in the United Kingdom, Perthshire in Scotland, Washington state in the United States, Vancouver Island in Canada,. In the Southern Hemisphere, tea is grown as far south as Hobart on the Australian island of Tasmania and Waikato in New Zealand.

Tea plants are propagated from seed and cuttings; about 4 to 12 years are needed for a plant to bear seed and about three years before a new plant is ready for harvesting. In addition to a zone 8 climate or warmer, tea plants require at least 127 cm (50 in) of rainfall a year and prefer acidic soils. Many high-quality tea plants are cultivated at elevations of up to 1,500 m (4,900 ft) above sea level. While at these heights the plants grow more slowly, they acquire a better flavour.

Two principal varieties are used: Camellia sinensis var. sinensis, which is used for most Chinese, Formosan and Japanese teas, and C. s. var. assamica, used in Pu-erh and most Indian teas (but not Darjeeling). Within these botanical varieties, many strains and modern clonal varieties are known. Leaf size is the chief criterion for the classification of tea plants, with three primary classifications being, Assam type, characterised by the largest leaves; China type, characterised by the smallest leaves; and Cambodian type, characterised by leaves of intermediate size.

A tea plant will grow into a tree of up to 16 m (52 ft) if left undisturbed, but cultivated plants are generally pruned to waist height for ease of plucking. Also, the short plants bear more new shoots which provide new and tender leaves and increase the quality of the tea.

Only the top 1–2 in of the mature plant are picked. These buds and leaves are called 'flushes'. A plant will grow a new flush every seven to 15 days during the growing season. Leaves that are slow in development tend to produce better-flavoured teas. Pests of tea include mosquito bugs of the genus Helopeltis (which are true bugs that must not be confused with the dipteran) that can tatter leaves, so they may be sprayed with insecticides.

ORIGIN AND HISTORY OF TEA

The history of tea is long and complex, spreading across multiple cultures over the span of thousands of years. Tea likely originated in southwest China during the Shang dynasty as a medicinal drink. An early credible record of tea drinking dates to the 3rd century AD, in a medical text written by Hua Tuo. Tea was first introduced to Portuguese priests and merchants in China during the 16th century. Drinking tea became popular in Britain during the 17th century. The British introduced tea production, as well as tea consumption, to India, in order to compete with the Chinese monopoly on tea.

GEOGRAPHIC ORIGINS

- "Camellia sinensis originated in southeast Asia, specifically around the intersection of latitude 29°N and longitude 98°E, the point of confluence of the lands of northeast India, north Burma, southwest China and Tibet. The plant was introduced to more than 52 countries, from this 'centre of origin'."

Based on morphological differences between the Assamese and Chinese varieties, botanists have long asserted a dual botanical origin for tea; however, statistical cluster analysis, the same chromosome number (2n=30), easy hybridization, and various types of intermediate hybrids and spontaneous polyploids all appear to demonstrate a single place of origin for Camellia sinensis — the area including the northern part of Burma, and Yunnan and Sichuan provinces of China.

Yunnan Province has also been identified as "the birthplace of tea...the first area where humans figured out that eating tea leaves or brewing a cup could be pleasant." Fengqing County in the Lincang City Prefecture of Yunnan Province in China is said to be home to the world's oldest cultivated tea tree, some 3,200 years old.

According to The Story of Tea, tea drinking likely began in modern-day Yunnan province during the Shang Dynasty (1500 BC–1046 BC), as a medicinal drink. From there, the drink spread to Sichuan, and it is believed that there "for the first time, people began to boil tea leaves for consumption into a concentrated liquid without the addition of other leaves or herbs, thereby using tea as a bitter yet stimulating drink, rather than as a medicinal concoction."

ORIGIN MYTHS

In one popular Chinese legend, Shennong, the legendary Emperor of China and inventor of agriculture and Chinese medicine was drinking a bowl of just boiled water due to a decree that his subjects must boil water before drinking it some time around 2737 BC when a few leaves were blown from a nearby tree into his water, changing the color. The emperor took a sip of the brew and was pleasantly surprised by its flavor and restorative properties. A variant of the legend tells that the emperor tested the medical properties of various herbs on himself, some of them poisonous, and found tea to work as an antidote. Shennong is also mentioned in Lu Yu's famous early work on the subject, The Classic of Tea. A similar Chinese legend goes that the god of agriculture would chew the leaves, stems, and roots of various plants to discover medicinal herbs. If he consumed a poisonous plant, he would chew tea leaves to counteract the poison. A rather gruesome legend dates back to the Tang Dynasty. In the legend, Bodhidharma, the founder of Chan Buddhism, accidentally fell asleep after meditating in front of a wall for nine years. He woke up in such disgust at his weakness that he cut off his own eyelids. They fell to the ground and took root, growing into tea bushes. Sometimes, another version of the story is told with Gautama Buddha in place of Bodhidharma.

Scholars however believe that tea drinking likely originated in the southwest of China, and that the Chinese words for tea themselves may have been originally derived from the Austro-Asiatic languages of the people who originally inhabited that area. Whether or not these legends have any basis in fact, tea has played a significant role in Asian culture for centuries as a staple beverage, a curative, and a status symbol. It is not surprising, therefore, that theories of its origin are often religious or royal in nature.

EARLY HISTORY

China

The Chinese have consumed tea for thousands of years. People of the Han Dynasty used tea as medicine (though the first use of tea as a stimulant is unknown). China is considered to have the earliest records of tea consumption, with possible records dating back to the 10th century BC. Note however that the current word for tea in Chinese only came into use in the 8th century AD, there are therefore uncertainties as to whether the older words used are the same as tea. The word tu appears in Shijing and other ancient texts to signify a kind of "bitter vegetable", and it is possible that it referred to a number of different plant, such as sowthistle, chicory, or smartweed, perhaps including tea. In the Chronicles of Huayang, it was recorded that the Ba people in Sichuan presented tu to the Zhou king. The state of Ba and its neighbour Shu were later conquered by the Qin, and according to the 17th century scholar Gu Yanwu who wrote in Ri Zhi Lu: "It was after the Qin had taken Shu that they learned how to drink tea."

The first known reference to boiling tea came from the Han dynasty work "The Contract for a Youth" written by Wang Bao where, among the tasks listed to be undertaken by the youth, "he shall boil tea and fill the utensils and "he shall buy tea at Wuyang". The first record of cultivation of tea also dated it to this period (Ganlu era of Emperor Xuan of Han) when tea was cultivated on

Meng Mountain near Chengdu. From the Tang to the Qing dynasties, the first 360 leaves of tea grown here were picked each spring and presented to the emperor. Even today its green and yellow teas, such as the Mengding Ganlu tea, are still sought after.

An unknown Chinese inventor was also the first person to invent a tea shredder. An early credible record of tea drinking dates to 220 AD, in a medical text Shi Lun by Hua Tuo, who stated, "to drink bitter t'u constantly makes one think better." Another possible early reference to tea is found in a letter written by the Qin Dynasty general Liu Kun. However, before the mid-8th century Tang dynasty, tea-drinking was primarily a southern Chinese practice. It became widely popular during the Tang Dynasty, when it was spread to Korea, Japan, and Vietnam.

Laozi, the classical Chinese philosopher, described tea as "the froth of the liquid jade" and named it an indispensable ingredient to the elixir of life. Legend has it that master Lao was saddened by society's moral decay and, sensing that the end of the dynasty was near, he journeyed westward to the unsettled territories, never to be seen again. While passing along the nation's border, he encountered and was offered tea by a customs inspector named Yin Hsi. Yin Hsi encouraged him to compile his teachings into a single book so that future generations might benefit from his wisdom. This then became known as the Dao De Jing, a collection of Laozi's sayings.

During the Sui Dynasty (589–618 AD) tea was introduced to Japan by Buddhist monks.

The Tang Dynasty writer Lu Yu's Cha Jing (The Classic of Tea) is an early work on the subject. According to Cha Jing tea drinking was widespread. The book describes how tea plants were grown, the leaves processed, and tea prepared as a beverage. It also describes how tea was evaluated. The book also discusses where the best tea leaves were produced. Teas produced in this period were mainly tea bricks which were often used as currency, especially further from the center of the empire where coins lost their value.

During the Song Dynasty (960–1279), production and preparation of all tea changed. The tea of Song included many loose-leaf styles (to preserve the delicate character favored by court society), but a new powdered form of tea emerged. Steaming tea leaves was the primary process used for centuries in the preparation of tea. After the transition from compressed tea to the powdered form, the production of tea for trade and distribution changed once again. The Chinese learned to process tea in a different way in the mid-13th century. Tea leaves were roasted and then crumbled rather than steamed. This is the origin of today's loose teas and the practice of brewed tea.

During the Yuan and Ming dynasties, unfermented tea leaves were first pan-fried, then rolled and dried. This stops the oxidation process which turns the leaves dark and allows tea to remain green. In the 15th century, Oolong tea, where the tea leaves were allowed to partially ferment before pan-frying, was developed.

Western taste, however, preferred the fully oxidized black tea, and the leaves were allowed to ferment further. Yellow tea was an accidental discovery in the production of green tea during the Ming dynasty, when apparently sloppy practices allowed the leaves to turn yellow, but yielded a different flavour as a result.

Tea production in China, historically, was a laborious process, conducted in distant and often poorly accessible regions. This led to the rise of many apocryphal stories and legends surrounding the harvesting process. For example, one story that has been told for many years is that of a village where monkeys pick tea. According to this legend, the villagers stand below the monkeys and taunt them. The monkeys, in turn, become angry, and grab handfuls of tea leaves and throw them at the villagers. There are products sold today that claim to be harvested in this manner, but no reliable commentators have observed this firsthand, and most doubt that it happened at all. For many hundreds of years the commercially used tea tree has been, in shape, more of a bush than a tree. "Monkey picked tea" is more likely a name of certain varieties than a description of how it was obtained. In 1391, the Ming court issued a decree that only loose tea would be accepted as a "tribute". As a result, loose tea production increased and processing techniques advanced. Soon, most tea was distributed in full-leaf, loose form and steeped in earthenware vessels.

Hong Kong

In Hong Kong, apart from the yum cha culture of southern China, a localised version of English tea was developed, the Hong Kong-style milk tea.

Japan

Tea use spread to Japan about the sixth century. Tea became a drink of the religious classes in Japan when Japanese priests and envoys, sent to China to learn about its culture, brought tea to Japan. Ancient recordings indicate the first batch of tea seeds were brought by a priest named Saichô (767–822) in 805 and then by another named Kûkai (774–835) in 806. It became a drink of the royal classes when Emperor Saga, the Japanese emperor, encouraged the growth of tea plants. Seeds were imported from China, and cultivation in Japan began.

In 1191, the famous Zen priest Eisai (1141–1215) brought back tea seeds to Kyoto. Some of the tea seeds were given to the priest Myoe Shonin, and became the basis for Uji tea. The oldest tea specialty book in Japan, Kissa Yôjôki (How to Stay Healthy by Drinking Tea), was written by Eisai. The two-volume book was written in 1211 after his second and last visit to China. The first sentence states, "Tea is the ultimate mental and medical remedy and has the ability to make one's life more full and complete." Eisai was also instrumental in introducing tea consumption to the warrior class, which rose to political prominence after the Heian Period.

Green tea became a staple among cultured people in Japan—a brew for the gentry and the Buddhist priesthood alike. Production grew and tea became increasingly accessible, though still a privilege enjoyed mostly by the upper classes.

The tea ceremony of Japan was introduced from China in the 15th century by Buddhists as a semi-religious social custom. The modern tea ceremony developed over several centuries by Zen Buddhist monks under the original guidance of the monk Sen no Rikyû (1522–1591). In fact, both the beverage and the ceremony surrounding it played a prominent role in feudal diplomacy.

In 1738, Soen Nagatani developed Japanese sencha, literally roasted tea, which is an unfermented form of green tea. It is the most popular form of tea in Japan today. In 1835, Kahei Yamamoto developed gyokuro, literally jewel dew, by shading tea trees during the weeks leading up to harvesting. At the end of the Meiji period (1868–1912), machine manufacturing of green tea was introduced and began replacing handmade tea.

Korea

The first historical record documenting the offering of tea to an ancestral god describes a rite in the year 661 in which a tea offering was made to the spirit of King Suro, the founder of the Geumgwan Gaya Kingdom (42–562). Records from the Goryeo Dynasty (918–1392) show that tea offerings were made in Buddhist temples to the spirits of revered monks.

During the Joseon Dynasty (1392–1910), the royal Yi family and the aristocracy used tea for simple rites. The "Day Tea Rite" was a common daytime ceremony, whereas the "Special Tea Rite" was reserved for specific occasions. Toward the end of the Joseon Dynasty, commoners joined the trend and used tea for ancestral rites, following the Chinese example based on Zhu Xi's text formalities of Family.Stoneware was common, ceramic more frequent, mostly made in provincial kilns, with porcelain rare, imperial porcelain with dragons the rarest.

The earliest kinds of tea used in tea ceremonies were heavily pressed cakes of black tea, the equivalent of aged pu-erh tea still popular in China. However, importation of tea plants by Buddhist monks brought a more delicate series of teas into Korea, and the tea ceremony. Green tea, "Jakseol" or "Jungno", is most often served. However, other teas such as "Byeoksoryeong" Cheonhachun, Ujeon, Okcheon, as well as native chrysanthemum tea, persimmon leaf tea, or mugwort tea may be served at different times of the year.

Vietnam

Vietnamese green teas have been largely unknown outside of mainland Asia until the present day. Recent free-enterprise initiatives are introducing these green teas to outside countries through new export activities. Some specialty Vietnamese teas include Lotus tea and Jasmine tea. Vietnam also produces black and oolong teas in lesser quantities. Vietnamese teas are produced in many areas that have been known for tea-house "retreats." For example, some are located amidst immense tea forests of the Lamdong highlands, where there is a community of ancient Ruong houses built at the end of the 18th century.

GLOBAL EXPANSION

The earliest record of tea in a more occidental writing is said to be found in the statement of an Arabian traveler, that after the year 879 the main sources of revenue in Canton were the duties on salt and tea. Marco Polo records the deposition of a Chinese minister of finance in 1285 for his arbitrary augmentation of the tea taxes.

The travelers Giovanni Batista Ramusio (1559), L. Almeida (1576), Maffei (1588), and Teixeira (1610) also mentioned tea. In 1557, Portugal established a trading port in Macau and word of the Chinese drink "chá" spread quickly, but there is no mention of them bringing any samples home.

In the early 17th century, a ship of the Dutch East India Company brought the first green tea leaves to Amsterdam from China. Tea was known in France by 1636. It enjoyed a brief period of popularity in Paris around 1648. The history of tea in Russia can also be traced back to the seventeenth century. Tea was first offered by China as a gift to Czar Michael I in 1618. The Russian ambassador tried the drink; he did not care for it and rejected the offer, delaying tea's Russian introduction by fifty years. In 1689, tea was regularly imported from China to Russia via a caravan of hundreds of camels traveling the year-long journey, making it a precious commodity at the time. Tea was appearing in German apothecaries by 1657 but never gained much esteem except in coastal areas such as Ostfriesland. Tea first appeared publicly in England during the 1650s, where it was introduced through coffeehouses. From there it was introduced to British colonies in America and elsewhere.

Portugal

Tea was first introduced to Portuguese priests and merchants in China during the 16th century, at which time it was termed chá.

Greece and Cyprus

Throughout Greece and Cyprus Greek tea is made with cinnamon and cloves.

India

Tea was first introduced into India by the British, in an attempt to break the Chinese monopoly on tea. The British, "using Chinese seeds, plus Chinese planting and cultivating techniques, launched a tea industry by offering land in Assam to any European who agreed to cultivate tea for export." Tea was originally only consumed by Anglicized Indians, it was not until the 1950s that tea grew widely popular in India through a successful advertising campaign by the India Tea Board.Prior to the British, the plant may have been used for medicinal purposes.

Some cite the Sanjeevani tea plant first recorded reference of tea use in India. However, scientific studies have shown that the Sanjeevani plant is in fact a different plant and is not related to tea. The Singpho tribe and the Khamti tribe also validate that they have been consuming tea since the 12th century. However, commercial production of tea in India did not begin until the arrival of the British East India Company, at which point large tracts of land were converted for mass tea production.

The Chinese variety is used for Sikkim, Darjeeling tea, and the Assamese variety, clonal to the native to Assam, everywhere else. The British started commercial tea plantations in India and in Ceylon: "In 1824 tea plants were discovered in the hills along the frontier between Burma and Assam. The British introduced tea culture into India in 1836 and into Ceylon (Sri Lanka) in 1867. At first they used seeds from China, but later seeds from the clonal Assam plant were used." Only black tea was produced until recent decades. India was the top producer of tea for nearly a century, but was displaced by China as the top tea producer in the 21st century. Indian tea companies have acquired a number of iconic foreign tea enterprises including British brands Tetley and Typhoo. While India is the largest consumer of tea worldwide, the per-capita consumption of tea in India remains a modest 750 grams per person every year. Recently consumption of green tea has seen a great upsurge across the cities. Average growth in the consumption is assumed to be over 50%. One estimate suggest the market size has already crossed over INR 1400crore and will reach a 6000 crore in next few years. Top station, 41 km (1 Hour) from Munnar, is aptly named, as it is home to some of the highest tea plantations in India. It lies on the state of Kerala and commands a panoramic view of rolling green hills.

Iran

Gilan in North of Iran is main production center of Iranian Tea. Historically, Lahijan is the first town in Iran to have tea plantations. With its mild weather, soil quality and fresh spring water, Lahijan stands to have the largest area of tea cultivation in Iran. "Lahijan Spring Tea" is the best quality tea produced in the country. Tea is cultivated at other cities of Gilan, for example Fuman and Roudsar.

Taiwan

Taiwan is famous for the making of Oolong tea and green tea, as well as many western-styled teas. Bubble Tea or "Zhen Zhu Nai Cha" is black tea mixed with sweetened condensed milk and tapioca. Since the island was known to Westerners for many centuries as Formosa — short for the Portuguese Ilha Formosa, or "beautiful island" — tea grown in Taiwan is often identified by that name.

Thailand

Thai tea or "cha-yen" in Thailand, is a drink made from strongly brewed black tea ("red tea" in East Asia). Other ingredients may include added orange blossom water, star anise, crushed tamarind seed or red and yellow food coloring, and sometimes other spices as well. This tea is sweetened with sugar and condensed milk.

Usually, Thai people drink Thai hot tea in the morning, frequently with Yau ja gwai (fried dough) or Pa-tong-ko.

The varieties of Thai tea include:

- Thai hot tea is Thai tea served hot.
- Dark Thai hot tea is Thai tea served hot with no milk content, sweetened with sugar only.
- Dark Thai iced tea is Thai tea served cold with ice and without milk.

Turkey

Turkey is traditionally one of the largest tea markets in the world. Turkish black tea is the most popular drink in Turkey, even more popular than Turkish coffee.

United Kingdom

The first record of tea in English came from a letter written by Richard Wickham, who ran an East India Company office in Japan, writing to a merchant in Macao requesting "the best sort of chaw" in 1615. Peter Mundy, a traveller and merchant who came across tea in Fujian in 1637, wrote, "chaa — only water with a kind of herb boyled in it ". In 1657, tea was listed as an item in the price list in a London coffee house, and the first advertisement for tea appear in 1658. On 25 September 1660 Samuel Pepys recorded in his diary: "I did send for a cup of tee (a China drink) of which I never had drank before." It is probable that early imports were smuggled via Amsterdam or through sailors arriving on eastern boats.

The marriage of King Charles II in 1662 to the Portuguese princess Catherine of Braganza also brought the tea drinking habit to court. Official trade of tea began in the 1660s at with an import of only 2 ounces, but grew to 24 million pounds a year by 1801.

Regular trade began in Canton (now Guangzhou), where it was controlled by two monopolies: the Chinese Cohong (trading companies) and the British East India Company. The Cohong acquired tea from 'tea men' who had an elaborate supply chain into the mountains and provinces where tea grew.

The East India Company brought back many products, of which tea was just one, but it was to prove one of the most successful. It was initially promoted as a medicinal beverage or tonic but by the end of the seventeenth century was taken as an all-purpose drink, albeit mainly by the elite as it was an expensive drink.

Tea was not traded in significant amount until the 18th century. Black tea overtook green tea in popularity in the 1720s and British drinkers added sugar and milk to tea, a practice that was not done in China. Tea became increasingly popular and by 1750 had become the national drink. The origin of large trade in tea was the need for a return cargo from the East Indies. Merchant ships delivered fabrics manufactured in Britain to India and China but would return empty or partially full. To solve this problem the East India Company began a vigorous public relations campaign in England to popularize tea among the common people in Britain and develop it as a viable return cargo. A fungus reduced Ceylon's coffee production by 95% in the 19th century, cementing tea's popularity.

The escalation of tea importation and sales over the period 1690 to 1750 is mirrored closely by the increase in importation and sales of cane sugar: the British were not drinking just tea but sweet tea. Thus, two of Britain's trading triangles converged: the sugar sourced from Britain's trading triangle encompassing Britain, Africa and the West Indies and the tea from the triangle encompassing Britain, India and China.

In China, the Qing dynasty Qianlong Emperor decreed that "China was the center of the world and had everything they could ever need, so all trade with foreigners must be paid for in silver!" This meant that British traders had to pay China for its tea with silver bullion. Critics of the tea trade at this time would point to the damage caused to Britain's wealth by this loss of bullion. As a way to generate the silver needed as payment for tea, Britain began exporting opium from the traditional growing regions of British India (in present-day Pakistan and Afghanistan) into China. Opium use in China had a long history however British importation of opium, which began in the late 18th century, increased fivefold between 1821 and 1837, and the Qing government attitude towards opium, which was often ambivalent, hardened as usage of the drug spread more widely across Chinese society. It finally began serious measures to curtail imports in 1838–39. Tea by now had become an important source of tax revenue for the British Empire and the banning of the opium trade and thus the creation of funding issues for tea importers was one of the main causes of the First Opium War.

Whilst waging war on China was one of Britain's tactics it also began to explore, then executed, a plan to use India for growing tea. After tea plants were smuggled out of China, plantations were established in areas such as Darjeeling, Assam, and Ceylon. As an attempt to circumvent its dependence on Chinese tea, the East India Company sent Scottish botanist Robert Fortune to China to purchase and bring out of China tea plants, which were then taken to India, although it was the discovery of native varieties of tea plant in India which proved more important for the development of production there. Tea remained a very important item in Britain's global trade, contributing in part to Britain's global dominance by the end of the eighteenth century. To this day tea is seen worldwide as a symbol of 'Britishness', but also, to some, as a symbol of old British colonialism. The London 2012 section of the paralympic handover in Beijing included tea as part of the routine. A cup or mug of tea in Britain is usually made in a different way than is common in China and other Eastern countries. Over 90% of tea consumed is black tea, often but not always with a small amount of milk and / or sugar added. The tea used is often contained in a tea bag. As of 2009 the UK can boast one commercial tea plantation with another planned. The existing one lies in Cornwall and is owned by the Tregothnan Estate. By 2015 another will lie in Pembrokeshire, Wales, owned by the Pembrokeshire Tea Company.

United States

While coffee is by far more popular, hot brewed black tea is enjoyed both with meals and as a refreshment by much of the population. Similarly, iced tea is consumed throughout. In the Southern states sweet tea, sweetened with large amounts of sugar or an artificial sweetener and chilled, is the fashion. Outside the South, sweet tea is sometimes found, but primarily because of cultural migration and commercialization. The drinking of tea in the United States was largely influenced by the passage of the Tea Act and its subsequent protest during the American Revolution. Tea consumption sharply decreased in America during and after the Revolution, when many Americans switched from drinking tea to drinking coffee, considering tea drinking to be unpatriotic. The American specialty tea market quadrupled in the years from 1993 to 2008, now being worth $6.8 billion a year. Similar to the trend of better coffee and better wines, this tremendous increase was partly due to consumers who choose to trade up. Specialty tea houses and retailers also started to pop up during this period.

Canada

Until the Second World War, Canadians were big tea drinkers since the days of British colonization. As a result of the war, Canadians drank more coffee like their American neighbors to the south. During the 1990s, Canadians begun to purchase more specialty teas instead of coffee. A commercial tea farm has opened on Vancouver Island in British Columbia in 2010 and expects production to start in 2015.

Australia

The Aboriginal Australians drank an infusion from the plant species leptospermum (a different plant from the tea plant or camellia sinensis). Upon discovering Australia, Captain Cook noticed the aboriginal peoples drinking it and called it tea. Today the plant is referred to as the "ti tree." Through colonisation by the British, tea was introduced to Australia. In fact, tea was aboard the First Fleet in 1788. Tea is a large part of modern Australian culture due to its British origins. Australians drink tea and have afternoon tea and morning tea much the way the British do. Additionally, due to Australia's climate, tea is able to be grown and produced in northern Australia. In 2000, Australia consumed 14,000 tonnes of tea annually. Tea production in Australia remains very small and is primarily in northern New South Wales and Queensland. Most tea produced in Australia is black tea, although there are small quantities of green tea produced in the Alpine Valleys region of Victoria.

In 1884, the Cutten brothers established the first commercial tea plantation in Australia in Bingil Bay in northern Queensland. In 1883, Alfred Bushell opened the first tea shop in Australia in present-day Queensland. In 1899, Bushell's sons moved the enterprise to Sydney and began selling tea commercially, founding Australia's first commercial tea seller Bushell's Company.

Sri Lanka

Sri Lanka is renowned for its high quality tea and as the fourth biggest tea producing country globally, after China, India and Kenya, and has a production share of 9% in the international sphere. The total extent of land under tea cultivation has been assessed at approximately 187,309 hectares. The plantations started by the British were initially taken over by the government in the 1960s, but have been privatized and are now run by 'plantation companies' which own a few 'estates' or tea plantations each. Ceylon tea is divided into 3 groups as Upcountry, Mid country and Low country tea based on the geography of the land on which it is grown.

Africa and South America

Africa and South America have seen greatly increased tea production in recent decades, the great majority for export to Europe and North America respectively, produced on large estates, often owned by tea companies from the export markets. Almost all production is of basic mass-market teas, processed by the Crush, Tear, Curl method. Kenya is now the third largest global producer, after China and India, and is now the largest exporter of tea to the United Kingdom. There is also a great consumption of tea in Chile. In South Africa, the non-Camellia sinensis beverage rooibos is popular. In South America yerba mate is a popular infused beverage. The only European plantation is Chá Gorreana, located in Ribeira Grande, São Miguel island, Azores (Portugal).

PROCESSING AND CLASSIFICATION OF TEA

Tea is generally divided into categories based on how it is processed. At least six different types are produced:

- *White:* Wilted and unoxidized
- *Yellow:* Unwilted and unoxidized, but allowed to yellow
- *Green:* Unwilted and unoxidized
- *Oolong:* Wilted, bruised, and partially oxidized
- *Black:* Wilted, sometimes crushed, and fully oxidized (called 'red tea' in China)
- *Post-Fermented:* Green tea that has been allowed to ferment/compost ('black tea' for the Chinese)

The most common are white, green, oolong, and black.

After picking, the leaves of C. sinensis soon begin to wilt and oxidize unless immediately dried. An enzymatic oxidation process triggered by the plant's intracellular enzymes causes the leaves to turn progressively darker as their chlorophyll breaks down and tannins are released. This darkening is stopped at a predetermined stage by heating, which deactivates the enzymes responsible. In the production of black teas, halting by heating is carried out simultaneously with drying. Without careful moisture and temperature control during manufacture and packaging, growth of undesired molds and bacteria may make tea unfit for consumption.

BLENDING AND ADDITIVES

Although single-estate teas are available, almost all tea in bags and most loose tea sold in the West is blended. Such teas may combine others from the same cultivation area or several different ones. The aim is to obtain consistency, better taste, higher price, or some combination of the three.

Tea easily retains odors, which can cause problems in processing, transportation, and storage. This same sensitivity also allows for special processing (such as tea infused with smoke during drying) and a wide range of scented and flavoured variants, such as bergamot (found in Earl Grey), vanilla, and spearmint.

NUTRIENTS AND PHYTOCHEMICALS

Black and green teas contain no essential nutrients in significant content, with the exception of the dietary mineral, manganese at 0.5 mg per cup or 26%

of the Daily Value. Tea leaves contain diverse polyphenols, including flavonoids, epigallocatechin gallate (commonly noted as EGCG) and other catechins. It has been suggested that green and black tea may protect against cancer or other diseases such as obesity or Alzheimer's disease, but the compounds found in green tea have not been conclusively demonstrated to have any effect on human diseases. One human study demonstrated that regular consumption of black tea over four weeks had no beneficial effect in lowering blood cholesterol levels.

TYEPS OF TEA AND CULTURE OF TEA

Tea may be consumed early in the day to heighten calm alertness; it contains L-theanine, theophylline, and bound caffeine (sometimes called theine). Decaffeinated brands are also sold. While herbal teas are also referred to as tea, most of them do not contain leaves from the tea plant.

While tea is the second most consumed beverage on Earth after water, in many cultures it is also consumed at elevated social events, such as afternoon tea and the tea party. Tea ceremonies have arisen in different cultures, such as the Chinese and Japanese tea ceremonies, each of which employs traditional techniques and ritualised protocol of brewing and serving tea for enjoyment in a refined setting.

One form of Chinese tea ceremony is the Gongfu tea ceremony, which typically uses small Yixing clay teapots and oolong tea. Turkish tea is an important part of Turkish cuisine, and is the most commonly consumed hot drink, despite the country's long history of coffee consumption. In 2004 Turkey produced 205,500 tonnes of tea (6.4% of the world's total tea production), which made it one of the largest tea markets in the world, with 120,000 tons being consumed in Turkey, and the rest being exported. In 2010 Turkey had the highest per capita consumption in the world at 2.7 kg. As of 2013, the per-capita consumption of Turkish tea exceeds 10 cups per day and 13.8 kg per year. Tea is grown mostly in Rize Province on the Black Sea coast.

Ireland has, for a long time, been one of the biggest per-capita consumers of tea in the world. The national average is four cups per person per day, with many people drinking six cups or more. Tea in Ireland is usually taken with milk or sugar and is slightly spicier and stronger than the traditional English blend. The two main brands of tea sold in Ireland are Lyons and Barry's. The Irish love of tea is perhaps best illustrated by the stereotypical housekeeper, Mrs. Doyle in the popular sitcom Father Ted.

Tea is prevalent in most cultures in the Middle East. In Arab culture, tea is a focal point for social gatherings. In Pakistan, tea is called chai. Both black and green teas are popular and are known locally as sabz chai and kahwah, respectively. The popular green tea called kahwah is often served after every meal in the Pashtun belt of Balochistan and in Khyber Pakhtunkhwa, which is where the Khyber Pass of the Silk Road is found. In the transnational Kashmir region, which straddles the border between India and Pakistan, Kashmiri chai or noon chai, a pink, creamy tea with pistachios, almonds, cardamom, and sometimes cinnamon, is consumed primarily at special occasions, weddings, and during the winter months when it is sold in many kiosks.

In central and southern Punjab and the metropolitan Sindh region of Pakistan, tea with milk and sugar (sometimes with pistachios, cardamom, etc.), commonly referred to as chai, is widely consumed. It is the most common beverage of households in the region. In the northern Pakistani regions of Chitral and Gilgit-Baltistan, a salty, buttered Tibetan-style tea is consumed. In Iranian culture, tea is so widely consumed, it is generally the first thing offered to a household guest.

In India, tea is one of the most popular hot beverages. It is consumed daily in almost all homes, offered to guests, consumed in high amounts in domestic and official surroundings, and is made with the addition of milk with or without spices. It is also served with biscuits dipped in the tea and eaten before consuming the tea. More often than not, it is drunk in "doses" of small cups (referred to as "Cutting" chai if sold at street tea vendors) rather than one large cup. On 21 April 2012, the Deputy Chairman of Planning Commission (India), Montek Singh Ahluwalia, said tea would be declared as national drink by April 2013. The move is expected to boost the tea industry in the country. Speaking on the occasion, Assam Chief Minister Tarun Gogoi said a special package for the tea industry would be announced in the future to ensure its development.

In the United States, 80% of tea is consumed as iced tea. Sweet tea is native to the southeastern US, and is iconic in its cuisine. Switzerland has its own unique blend of iced tea, made with the basic ingredients like black tea, sugar, lemon juice and mint, but a variety of Alp herbs are also added to the concoction. Apart from classic flavours like lemon and peach, exotic flavours like jasmine and lemongrass are also very popular.

In the United Kingdom, it is consumed daily and often by a majority of people across the country, and indeed is perceived as one of Britain's cultural beverages. In British homes, it is customary good manners for a host to offer tea to guests soon after their arrival. Tea is generally consumed at home; outside the home in cafés. Afternoon tea with cakes on fine porcelain is a cultural stereotype, sometimes available in quaint tea-houses. In southwest England, many cafes serve a 'cream tea', consisting of scones, clotted cream, and jam alongside a pot of tea. Throughout the UK, 'tea' may also refer to the evening meal. In Burma (Myanmar), tea is consumed not only as hot drinks, but also as sweet tea and green tea known locally as laphet-yay and laphet-yay-gyan, respectively. Pickled tea leaves, known locally as laphet, are also a national delicacy. Pickled tea is usually eaten with roasted sesame seeds, crispy fried beans, roasted peanuts and fried garlic chips.

BUBBLE TEA

Bubble tea (also known as pearl milk tea, or boba-milk tea) is a Taiwanese tea-based drink invented in Taichung in the 1980s. Most bubble tea recipes contain a tea base mixed/shaken with fruit or milk, to which chewy tapioca balls and/or fruit jellies are often added. The "bubble" refers to the foam created by shaking the tea, which itself is called "pàomò" tea, meaning "frothy/foamy" tea. Ice-blended versions are usually mixed with fruit or syrup, resulting in a slushy consistency.

There are many variants of the drinks, and many kinds of ingredients may be added. The most popular bubble drinks are bubble milk tea with tapioca and bubble milk green tea with tapioca.

Bubble teas are typically of two distinct types: fruit-flavored teas and milk teas. However, some shops offer hybrid "fruit milk teas". Most milk teas include powdered dairy or non-dairy creamers, but some shops also offer fresh milk as an alternative. Other varieties are 100% crushed-fruit smoothies with tapioca pearls and signature ice cream shakes made from local ice cream sources. Many American bubble tea vendors sell "milk smoothies", which are similar to bubble tea but do not contain any tea ingredients. Some small cafés offer sweetener substitutes, such as honey, agave, stevia, and aspartame, upon special request.

The oldest known bubble tea consisted of a mixture of hot Taiwanese black tea, small tapioca pearls, condensed milk, and syrup or honey. Many variations were created, the most common of which is served cold rather than hot. The tea type is frequently replaced. First was bubble green tea, which uses jasmine-infused green tea instead of black tea. Big tapioca pearls were adapted and quickly replaced the small pearls. Peach or plum flavoring appeared, then more fruit flavors were added until, in some variations, the tea was removed entirely in favor of real fruit.

These fruit versions sometimes contain colored pearls (and/or "jelly cubes" as in the related drink taho), the color chosen to match whatever fruit juice is used. Flavors may be added in the form of powder, fruit juice, pulp, or syrup to hot black or green tea, which is then shaken in a cocktail shaker or mixed with ice in a blender. Cooked tapioca pearls and other mix-ins (such as vanilla extract, honey, syrup, and sugar) are added at the end.

Today, one can find shops entirely devoted to bubble tea, similar to the juice bars of the early 1990s. Some cafés use plastic dome-shaped lids, while other bubble tea bars serve it using a machine to seal the top of the cup with plastic cellophane.

The latter method allows the tea to be shaken in the serving cup and makes it spill-free until one is ready to drink it. The cellophane is then pierced with an oversized straw large enough to allow the pearls to pass through. Today, in Taiwan, it's more common for people to refer to the drink as "pearl milk tea" ("zhen zhû nai chá", or "zhen nai" for short). "Pearl milk tea" is also used by English speakers and overseas Chinese and Taiwanese speakers, but it is usually called "bubble tea" or "boba tea" by English speakers, with the former seemingly more common in locations with less Chinese influence.

VARIANTS

Each of the ingredients of bubble tea can have many variations depending on the tea store. Typically, different types of black tea, green tea, or even coffee form the basis of this beverage. The most common black tea varieties are oolong and Earl Grey, while jasmine green tea is a mainstay at almost all tea stores. Another variation called yuanyang originated in Hong Kong and consists of half black tea and half coffee. Some people add milk to the drink. Decaffeinated versions of teas are sometimes available when the tea house freshly brews the tea base.

The milk in bubble tea is optional, though many tea stores use it. Some cafés use a non-dairy creamer milk substitute instead of milk because many East Asians are lactose intolerant and because it is cheaper and easier to store and use than perishable milk. In Western countries, soy milk options are widely available for those who avoid dairy products. This adds a distinct flavor and consistency to the drink.

Different flavorings can be added to bubble tea. Some widely available fruit flavors include strawberry, green apple, passion fruit, mango, lemon, watermelon, grape, lychee, peach, pineapple, cantaloupe, honeydew, banana, avocado, coconut, kiwifruit, and jackfruit. Other popular non-fruit flavors include taro, pudding, chocolate, coffee, mocha, barley, sesame, almond, ginger, lavender, rose, caramel, and violet. Some of the sour fruit flavors are available in bubble tea without milk only as the acidity tends to curdle the milk.

Other varieties of the bubble tea drink can include blended drinks. Many stores in the US provide a list of choices to choose from. Some may include coffee-blended drinks or even smoothies. Tapioca balls are the prevailing chewy tidbits in bubble tea, but a wide range of other options can be used to add similar texture to the drink. Green pearls have a small hint of green tea flavor and are chewier than the traditional tapioca balls. Jelly is also used in small cubes, stars, or rectangular strips, with flavors such as coconut jelly, konjac, lychee, grass, mango, and green tea often available at some shops. Rainbow, a fruit mix of konjac, has a pliant, almost crispy consistency. Azuki bean or mung bean paste, also typical toppings for Taiwanese shaved ice desserts, give the drinks an added subtle flavor as well as texture. Aloe, egg pudding (custard pudding), sago, and taro balls can also be found in most tea houses.

Due to its popularity, single-serving packets of black tea (with powdered milk and sugar included) are available as "Instant Boba Milk Tea" in some places.

Bubble tea cafés will also frequently serve drinks without coffee or tea in them. The base for these drinks is flavoring blended with ice, often called Snow Bubble. All mix-ins that can be added to the bubble tea can also be added to these slushie-like drinks. One drawback to them is that the coldness of the iced drink may cause the tapioca balls to harden, making them difficult to suck up through a straw and chew. To prevent this from happening, these slushies must be consumed more quickly than bubble tea.

Occasionally, nata de coco is used in mass-produced bubble tea drinks as a healthier alternative to tapioca starch. Nata de coco is high in dietary fiber and low in cholesterol and fat. The nata de coco is sliced into thin strips to make it easier to pass through a straw.

TYPES

- Pàomò hóngchá: "foam red tea", by direct translation, is the drink that is more appropriate for the more literal name of "bubble tea"; however, the English name, foam tea, is not used much throughout Asia. Consequently, in non-Chinese-speaking Asian countries, "bubble tea" is commonly used to refer to this drink. There is no tapioca in this particular drink. To create this, vendors mix hot or warm tea (in this case, black tea) with syrup or sugar and ice cubes into a cocktail shaker. Then they would shake the shaker either by hand or by machine before it is served. The resulting tea would be covered by a layer of foam or froth and the tea would have a light foamy feel to the taste.
- Pàomò naichá: "foam milk tea". One of the many variants that is prepared the same way as the "foam red tea", well-shaken before serving.
- Zhenzhû naichá: "pearl milk tea", or more commonly referred to as bubble tea by most English speakers and overseas Chinese speakers. The "pearl" name originally referred to the small 1/12" tapioca pearls added to the drink. Though most modern vendors serve only the bigger 1/4" pearls, they still use "pearl tea" as the name.
- Bôbà naichá: "bubble milk tea" and also commonly referred to as boba tea by English speakers and Asian Americans. The name refers to the variant with the bigger, 1/4" tapioca pearls.
- Hei zhenzhû naichá: "black pearl milk tea". Since the bigger 1/4" tapioca pearls are separately sold as "black pearls" in markets, this name was the logical first choice and is more popular with the consumers of the drink. "Boba" is an alternative name that is less commonly used these days.
- Chá zhenzhû): " tea pearl" .
- Pào pào chá: used interchangeably with refer to "bubble tea" in Singapore.

HEALTH CONCERNS

Tapioca pearls, milk powder, and juice syrups have in the past been found to contain banned chemical additives. In May 2011, a food scandal broke out in Taiwan where DEHP (a chemical plasticizer and potential carcinogen used to make plastic) was found as a stabilizer in drinks and juice syrups. Some of these products may have been exported and used in bubble tea shops around the world. DEHP can affect hormone balances. In June 2011, the Health Minister of Malaysia, Liow Tiong Lai, instructed companies selling "Strawberry Syrup", a material used in some bubble teas, to stop selling them after chemical tests showed they were tainted with a carcinogen identified as DEHP.

In August 2012, scientists from the Technical University of Aachen in Germany analyzed bubble tea samples within a research project to look for allergenic substances.

The result indicated that the products contain styrene, acetophenone and brominated substances. The report was published by German newspaper Rheinische Post and caused Taiwan's representative office in Germany to issue a statement, saying food items in Taiwan are monitored. Taiwan's Food and Drug Administration confirmed in September that in a second round of tests conducted by German authorities, Taiwanese bubble tea was found to be free of cancer-causing chemicals. The products were also found to contain no excessive levels of heavy-metal contaminants or other health-threatening agents.

In May 2013 the Taiwan Food and Drug Administration issued an alert on the detection of maleic acid, an unapproved food additive, in some food products, including tapioca pearls. The Agri-Food & Veterinary Authority of Singapore conducted its own tests and found additional brands of tapioca pearls and some other starch-based products sold in Singapore were similarly affected.

In April 2015 investigation for cleaning a pitcher containing the alleged poisonous milk tea at the store namely ErgoCHA in Sampaloc, Manila Philippines denied the allegation of their helper that he brought a chemical to the store the night before the alleged poisoning happened. However, preliminary findings of the Department of Health and Food and Drug Administration did not find any toxic substance in the sample milk tea. The DOH and FDA are still conducting further tests to determine what caused the deaths. In a news report, ABS-CBN News told that the pitcher containing the alleged poisonous milk tea might be containing a cyanide.

As of May 10, 2015, ABS-CBN's news program TV Patrol reported that the victims of the milk tea poisoning at ErgoCha in Sampaloc, Manila was contaminated by a cleaning agent, oxalic acid. Although, investigation is still on-going for further information about the said incident.

HISTORY

The most accredited story about the creation of bubble tea is the Chun Shui Tang Teahouse in Taichung, Taiwan. The founder of the teahouse, Liu Han-Chieh, observed how the Japanese served cold coffee while on a visit in the 1980s, and applied this method to tea.

The new style of serving tea propelled his business, and multiple chains were established. This expansion would be the stepping stone for the rapid expansion of bubble tea. The actual creator of bubble tea is Lin Hsiu Hui, the teahouse's product development manager, who poured a sweetened pudding with tapioca balls into the iced tea drink during a meeting in 1988. The beverage was well received by the people at the meeting, leading to its inclusion on the menu.

It ultimately became the franchise's top-selling product. An alternative origin is the Hanlin teahouse in Tainan, Taiwan, owned by Tu Tsong-he. He made tea using traditional white tapioca, which has the appearance of pearls, supposedly resulting in the so-called "pearl tea". Shortly after, Hanlin changed the white tapioca balls to the black version that is seen most today.

The drink became popular in most parts of East and Southeast Asia during the 1990s. Famous chain-stores that sell boba teas are Gong Cha, ShareTea, and Coco The drink is well received by global consumers from Canada and USA, specifically around areas with high Asian demographics.

BLACK DRINK

The black drink is a name for several kinds of ritual beverages brewed by Native Americans in the Southeastern United States. Traditional ceremonial people of the Caddo, Cherokee, Choctaw, Muscogee and some other Indigenous peoples of the Southeastern Woodlands use the black drink in purification ceremonies. It was occasionally known as white drink because of the association of the color white with peace leaders in some Native cultures in the Southeast.

The black drink is typically seen in Native cultures where it is present as a purifier that removes spiritual and physical contamination from the drinker, and as such is never taken casually. While the recipe can vary between tribes, and the full formula is not published or given to outsiders, a prominent ingredient in many formulas is the roasted leaves and stems of Ilex vomitoria native to the Atlantic and Gulf Coasts.

The black drink usually contains emetic herbs, which are used with the express purpose of inducing vomiting. There is a degree of variation in its ritual use. It often involves vomiting, but not always. Vomiting is not induced by the mere consumption of a drink prepared with Ilex vomitoria. It was occasionally used by some groups as a stimulating, but not casual social drink for before council meetings.

PREPARATION

According to the ethnohistorical record, the yaupon leaves and branches used for the black drink were traditionally picked as close to the time of its planned consumption as possible. After picking they were lightly parched in a ceramic container over fire. The roasting increases the solubility in water of the caffeine, which is the same reason coffee beans are roasted. After browning, they were boiled in large containers of water until the liquid reached a dark brown or black color, giving it its name.

The liquid was then strained into containers to cool, until it was cool enough to not scald the skin, and drunk while still hot. Because caffeine is 30 times more soluble in boiling water than room temperature water, this heightened its effect. It was then consumed in a ritual manner. Its physiological effects are mainly those of massive doses of caffeine.

Three to six cups of strong coffee is equal to 0.5 to 1.0 grams of caffeine; the black drink could have delivered at least this much and possibly up to 3.0 to 6.0 grams of caffeine. Similar methods of production were adopted by European colonists for the production of a drink that often shared the same names with Native names for the black drink, but were used for different purposes.

While the general method of production is known, not all details of the preparation and ceremonial usage of the black drink are.The source of the emetic effect of black drink is not known and has been speculated upon by historians, archaeologists, and botanists. Some professionals belief it be caused by the addition of poisonous Button snakeroot. It has also been speculated that the emetic effect was caused by physically triggering the gag reflex or by the mere act of consuming vast quantities of a warm liquid.

Modern preparation and usage the black drink by Native Americans is less well documented. Online recipes for the black drink have been criticized by some Native Americans as potentially dangerous if not poisonous due those recipes leaving out key steps. The berries of the yaupon holly are poisonous. Kidney failure is one possible outcome of consuming beverages containing holly leaves. Those recipes have also been seen as violating Indigenous intellectual property and potentially leading to cultural appropriation. Some Native Americans warn that recipes found online or in books should not be assumed to be safe.

PRECONTACT USE

Archaeologists have demonstrated the use of various kinds of black drink among Native American groups stretching back far into antiquity, possibly dating to Late Archaic times. During the Hopewell period, the shell cups known from later black drink rituals become common in high status burials along with mortuary pottery and engraved stone and copper tablets. The significance of the shell cups may indicate the beginning of black drink ceremonialism. The fact that both the shells and the yaupon holly come from the same geographical location may indicate they were traded together in the Hopewell Interaction Sphere. The appearance of shell cups can be used as a virtual marker for the advent of Hopewell Culture in many instances. During the Mississippian culture period, the presence of items associated with the black drink ceremony had spread over most of the south, with many hundreds of examples from Etowah, Spiro, Moundville and Hiwassee Island.

Black drink at Cahokia

Archaeologists working at Cahokia, the largest Mississippian culture settlement located near the modern city of St. Louis, found distinctive and relatively rare pottery beakers dating from 1050 to 1250 CE. The beakers are small round pots with a handle on one side and a tiny lip on the opposing side. The surfaces of the unfired vessels was incised with motifs representing water and the underworld and resemble the whelk shells known to have been used for the consumption of the beverage during historic times. The inside of the vessels were found to be coated with a plant residue, which when tested was found to contain theobromine, caffeine and ursolic acid in the right proportions to have come from the Ilex vomitoria. The presence of the black drink in the Greater Cahokia area at this early date pushes back definitive use of the black drink by several centuries. The presence of the black drink hundreds of miles outside of its natural range on the East and Gulf coasts is evidence of a substantial trade network with the southeast, a trade that also involved sharks teeth and whelk shells.

Shell Cups

In historic accounts from the 16th and 17th century, black drink is usually imbibed in rituals using a cup made of marine shell. Three main species of marine shells have been identified as being used as cups for the black drink, lightning whelk, emperor helmet, and the horse conch. The most common was the lightning whelk, which has a left-handed or sinistral spiral. The left-handed spiral may have held religious significance because of its association with dance and ritual.

The center columnella, which runs longitudinally down the shell, would be removed, and the rough edges sanded down to make a dipper like cup. The columnella would then be used as a pendant, a motif that shows up frequently in Southeastern Ceremonial Complex designs. In the archaeological record columnella pendants are usually found in conjunction with bi-lobed arrows, stone maces, earspools, and necklace beads(all of which are motifs identified with the falcon dancer/warrior/chunkey player mythological figure). Artifacts made from these marine shells have been found as far north as Wisconsin and as far west as

Oklahoma. Several examples of cups from Moundville and Spiro have been found to have rings of black residue in the bottoms, suggesting they were used for black drink rituals. Many examples of shell cups found in Mississippian culture mounds are engraved with S.E.C.C. imagery. A few examples portray what is theorized to be black drink rituals, including what some anthropologists have interpreted as vomit issuing from the mouths of mythological beings.

POST CONTACT USE

A number of tribes across the Southeastern United States use a form of the black drink in their ceremonies. Muscogee Creeks, Cherokees, Choctaws, Ais, Guale, Chickasaws, Chitimacha, Timucua and others are documented users of a type of black drink. Although rituals vary amongst the different tribes, there are some common traits among them. Black drink is most commonly drunk by men, who also prepare it, but there are some accounts of its preparation and use by women among some groups. Removal of tribes to areas outside the natural range of Ilex vomitoria has been partially responsible for a decline in the preparation of the black drink among present Native Americans.

Ais

In 1696, Jonathan Dickinson witnessed the use of a beverage brewed from the Yaupon Holly among the Ais of Florida. Dickinson later learned that the Spanish called the plant casseena. The Ais parched the leaves in a pot, and then boiled them. The resulting liquid was then transferred to a large bowl using a gourd that had a long neck with a small hole at the top, and a 2-inch-wide hole in the side. On the occasion Dickinson witnessed, he estimated that there was nearly three gallons of the beverage in the bowl. After the liquid had cooled, the chief was presented with a conch-shell of the beverage. The chief threw part of it on the ground as a blessing and drank the rest. The chief's associates were then served in turn. Lower status men, women and children were not allowed to touch or taste the beverage. The chief and his associates sat drinking this brew, smoking and talking for most of the day. In the evening the bowl that had held the beverage was covered with a skin to make a drum. The Ais, accompanied by the drum and some rattles, sang and danced until the middle of the night.

Cherokee

Cherokee black drink is taken for purification at several traditional ceremonies. Made with emetics, the complete recipe is not shared with the public. The black drink induces vomiting for purification purposes. Other ritual medicinal beverages are also used in the ceremonies. Ilex vomitoria was not native to the traditional area controlled by the Cherokee before removal, but it was brought in to the area through trade and then transplanted.

Muscogean Peoples

Among the Muscogee the blank drink is called ássi. In the ceremonies of some cultures that use the drink, after its preparation it is passed out to the highest status person first, then the next highest status, and so forth. During each persons turn to drink, ritual songs may be sung Yahola. Its use was traditionally limited to only adult men. The ritual name Asi Yahola or Black Drink Singer is corrupted into English as Osceola).

Timucua

Among the Timucua, a type of black drink was called cacina by the Spanish. Preparation and consumption of the drink was strictly limited to the community council house. Women (other than an occasional female chief) were normally excluded from the council house except for activities such as dances, but did prepare the cacina. In 1678 a bedridden caçica was given permission to brew and consume cacina in her house, on the condition that no one else could be present while she did so.

CASSINA OR YAUPON TEA

After European contact with tribes in what is today the Southeastern United States, colonists began using the charred leaves of the yaupon holly to make a tea similar to the black drink, but without the ritual vomiting of it. Its use by colonists in Spanish Florida is documented as far back as 1615. An account from that year describes Spaniards as experiencing symptoms that would now be described as caffeine dependence due to daily consumption of what they called cacina or té del indio.

The use of Ilex vomitoria by colonists for tea making and for medicinal uses in the Carolinas is documented by the early eighteenth century. In the English speaking colonies it was known variously as cassina, yaupon tea, Indian tea, Carolina tea, and Appalachian tea. It was commonly believed to be and used as a diuretic.

By the late 1700s, yaupon tea was described as being more commonly used in North Carolina at breakfast than tea made with Camellia sinensis. In addition to using it on their own, Spanish and British colonist often drank black drink when engaging in discussions and treaties with Natives. Its preparation by British and Spanish colonists was nearly identical to the method of preparation used by their Native neighbors. Its use by colonists in French Louisiana is speculated to have occurred, but lacks documentation other than one source describing its medicinal uses from 1716.

During the Civil War, yaupon tea was used as a substitute for coffee and tea throughout the South. Yaupon continued to be used in North Carolina for medicinal purposes and as a common drink until the late 1890s. At that point, its use was stigmatized because of its natural abundance as being a habit associated with rural poor people. By 1928 it was described as only being in common use by on Knotts Island, North Carolina. During the Interwar period the United States Department of Agriculture investigated the use of cassina tea as a substitute for coffee and tea. There were also a few attempts at the commercialization of cassina tea during that same period. By 1973 it was believed that cassina tea was only being served at the Pony Island Restaurant on Orcracoke Island, North Carolina. In the early 2000s yaupon tea began witnessing a resurgence in its popularity, and can now be purchased online and at several historical sites related to Native Americans.

BUTTER TEA

Butter tea, also known as po cha, cha süma, Mandarin Chinese or gur gur in local Ladakhi terms, is a drink of the people in the Himalayan regions of Nepal, India and Bhutan and most famously, Tibet. It is also consumed in Bhutan and by Buddhist minorities in India. Traditionally, it is made from tea leaves, yak butter, water, and salt, however, given its wider availability and reduced cost, butter made from cow's milk is increasingly used. Drinking butter tea is a regular part of Tibetan life. Before work, a Tibetan will typically enjoy several bowlfuls of this beverage, and it is always served to guests. Nomads are said to often drink up to 40 cups of it a day. Since butter is the main ingredient, butter tea provides plenty of caloric energy and is particularly suited to high altitudes. The butter may also help prevent chapped lips.

According to the Tibetan custom, butter tea is drunk in separate sips, and after each sip the host refills the bowl to the brim. Thus, the guest never drains his bowl; rather, it is constantly topped up. If the visitor does not wish to drink, the best thing to do is leave the tea untouched until the time comes to leave and then drain the bowl. In this way etiquette is observed and the host will not be offended. Butter tea is also used for eating tsampa by pouring onto it, or dipping the tsampa into it, and mixing well. The concentrate, produced by repeatedly boiling tea leaves, will keep for several days, and is commonly used in towns. The tea is then combined with salt and butter in a special tea churn and churned vigorously before serving hot. Now an electric blender is often used.

HISTORY

While evidence of tea has been found in Tibet from before the 10th century, it did not reach its nearly universal status until about the 13th century, the time of the Sakya hierarchy and the Phagmodu kings. By the start of the rule by the Dalai Lama, tea had become a government monopoly. While it is now officially allowed to be sold by anyone, it still is mostly sold by government officials.

PREPARATION

The highest quality tea is made by boiling the tea leaves in water for half a day, achieving a dark brown color. It is then skimmed, and poured into a cylinder with fresh yak butter and salt which is then shaken. The result is a purplish liquid that is about the thickness of a stew or thick oil. It is then poured into clay tea-pots, or jars, that resemble Japanese teapots. Another method is to boil water, and add handfuls of the tea into the water, which is allowed to steep until it turns almost black. Salt is then added, along with a little soda if wanted. The tea is then strained through a horse-hair or reed colander into a wooden butter churn, and a large lump of butter is added. This is then churned until the tea reaches the proper consistency and transferred to copper pots that sit on a brazier to keep them warm. When a churn is not available, a wooden bowl and rapid stirring will suffice. Nowadays, when tea leaves, yak butter and wooden butter churn is not available, people often make butter tea using tea bags, different types of butter available in the market and a blender to churn.

PREPARATION OF STEEPING TEA

The traditional method of preparing tea is to place loose tea leaves directly (or in a tea infuser) into a tea pot or teacup, pour freshly boiled water over the leaves, and allow the infused liquid to steep (or "brew"). After a few minutes, the infuser is removed, or the tea is poured through a strainer while serving. Strength should be varied by the amount of tea leaves used, not changing the steeping time. Most green teas should be allowed two or three minutes, although other types may vary between thirty seconds and ten minutes. Quantity also varies by tea type, with a basic recipe calling for one slightly heaped teaspoon (about 5 ml) for each teacup of water (200–240 ml) (7–8 oz). Stronger teas to be drunk with milk (such as Assam) are often prepared more heavily, while more delicate high-grown varieties (such as a Darjeeling) more lightly. Optimum brewing temperature depends on tea type. Camellia sinensis naturally contains tannins having bitter properties accentuated by both temperature and steeping time. These tannins are enhanced by oxidation during processing. Teas with little or no oxidation, such as a green or white, are best at lower temperatures between 65 and 85 °C (149 and 185 °F), while more oxidized teas require 100 °C (212 °F) to extract their large, complex, flavourful phenolic molecules. In addition, boiling reduces the dissolved oxygen content of water, which would otherwise react with phenolic molecules to degrade them.

Type	Water temp.	Steep time	Infusions
White tea	65 to 70 °C (149 to 158 °F)	1–2 minutes	3
Yellow tea	70 to 75 °C (158 to 167 °F)	1–2 minutes	3
Green tea	75 to 80 °C (167 to 176 °F)	1–2 minutes	4–6
Oolong tea	80 to 85 °C (176 to 185 °F)	2–3 minutes	4–6

Black tea	99 °C (210 °F)	2–3 minutes	2–3
Flowering tea	100 °C (212 °F)	2–3 minutes	4–5
Pu'er tea	95 to 100 °C (203 to 212 °F)	Limitless	Several
Tisanes	99 °C (210 °F)	3–6 minutes	Varied

Some tea sorts are often brewed several times using the same leaves. Historically in China, tea is divided into a number of infusions. The first is immediately poured out to wash the tea, and then the second and further infusions are drunk. The third through fifth are nearly always considered the best, although different teas open up differently and may require more infusions to produce the best flavour. One way to taste a tea throughout its entire process is to add hot water to a cup containing the leaves and sample it every 30 seconds. As the tea leaves unfold (known as "The Agony of the Leaves") the taste evolves. A tea cosy or a teapot warmer are often used to keep the temperature of the tea in a teapot constant over periods of 20–60 minutes.

COLD BREW TEA (AKA REFRIGERATOR TEA) AND SUN TEA

While most tea is prepared using hot water, it is also possible to brew a beverage from tea using room temperature or cooled water. This requires longer steeping time to extract the key components, and produces a different flavor profile. For best results, it is best to use about 1.5 times the tea leaves that would be used for hot steeping, and to refrigerate for 4-10 hours. The process of making cold brew tea is much simpler than that for cold brew coffee. Cold brewing has some disadvantages compared to hot steeping.

Firstly, if the leaves or source water contain unwanted bacteria, they may flourish, whereas using hot water has the benefit of killing most bacteria. This is less of a concern in modern times and developed regions. Secondly, cold brewing may allow for less caffeine to be extracted, which may or may not be desired. Sun Tea is made by steeping the tea leaves in a jar of unheated tap water left in the sun. It does not get hot enough to kill bacteria present on the tea leaves or in the water, such as Alcaligenes viscolactis.

BLACK TEA

Popular varieties of black tea include Assam, Iran, Nepal, Darjeeling, Nilgiri, Turkish, Keemun, and Ceylon teas. Many of the active substances in black tea do not develop at temperatures lower than 90 °C (194 °F). As a result, black tea in the West is usually steeped in water near its boiling point, at around 99 °C (210 °F). The most common fault when making black tea is to use water at too low a temperature. Since boiling point drops with increasing altitude, it is difficult to brew black tea properly in mountainous areas. Warming the tea pot before steeping is critical at any elevation.

Western black teas are usually brewed for about four minutes and are usually not allowed to steep for less than 30 seconds or more than about five minutes (a process known as brewing or mashing in Britain). In many regions of the world, however, actively boiling water is used and the tea is often stewed. In India, black tea is often boiled for fifteen minutes or longer to make Masala chai, as a strong brew is preferred. Tea should be strained while serving. A food safety management group of the International Organization for Standardization (ISO) has published a standard for preparing a cup of tea (ISO 3103: Tea — Preparation of liquor for use in sensory tests), primarily intended for standardizing preparation for comparison and rating purposes.

GREEN TEA

In regions of the world that prefer mild beverages, such as the West and Far East, green tea should be steeped in water around 80 to 85 °C (176 to 185 °F), the higher the quality of the leaves the lower the temperature. Regions such as North Africa or Central Asia prefer a bitter tea, and hotter water is used. In Morocco, green tea is steeped in boiling water for 15 minutes.

The container in which green tea is steeped is often warmed beforehand to prevent premature cooling. High-quality green and white teas can have new water added as many as five or more times, depending on variety, at increasingly higher temperatures.

FLOWERING TEA

Flowering tea or blooming tea should be brewed at 100 °C (212 °F) in clear glass tea wares for up to three minutes. First pull 1/3 water to make the tea ball wet and after 30 seconds add the boiling water up to 4/5 of the capacity of the tea ware. The boiling water can help the tea ball bloom quickly and with a strong aroma of the tea. The height of glass tea ware should be 8–10 cm, which can help the tea and flowers bloom completely. One tea ball can be brewed 4-5 times.

OOLONG TEA

Oolong tea should be brewed around 185 to 205 °F, with the brewing vessel warmed before pouring the water. Yixing purple clay teapots are the traditional brewing-vessel for oolong tea which can be brewed multiple times from the same leaves, unlike green tea, seeming to improve with reuse. In the Chinese and Taiwanese Gongfu tea ceremony, the first brew is discarded, as it is considered a rinse of leaves rather than a proper brew.

PREMIUM OR DELICATE TEA

Some teas, especially green teas and delicate oolong teas, are steeped for shorter periods, sometimes less than 30 seconds. Using a tea strainer separates the leaves from the water at the end of the brewing time if a tea bag is not being used. However, the black Darjeeling tea, a premium Indian tea, needs a longer than average steeping time. Elevation and time of harvest offer varying taste profiles; proper storage and water quality also have a large impact on taste.

PU-ERH TEA

Pu-erh teas require boiling water for infusion. Some prefer to quickly rinse pu-erh for several seconds with boiling water to remove tea dust which accumulates from the ageing process, then infuse it at the boiling point (100 °C or 212 °F), and allow it to steep from 30 seconds to five minutes.

SERVING

To preserve the pretannin tea without requiring it all to be poured into cups, a second teapot may be used. The steeping pot is best unglazed earthenware; Yixing pots are the best known of these, famed for the high-quality clay from which they are made. The serving pot is generally porcelain, which retains the heat better. Larger teapots are a post-19th century invention, as tea before this time was very rare and very expensive. Experienced tea-drinkers often insist the tea should not be stirred around while it is steeping (sometimes called winding or mashing in the UK). This, they say, will do little to strengthen the tea, but is likely to bring the tannins out in the same way that brewing too long will do. For the same reason, one should not squeeze the last drops out of a teabag; if stronger tea is desired, more tea leaves should be used.

ADDITIVES

The addition of milk to tea in Europe was first mentioned in 1680 by the epistolist Madame de Sévigné. Many teas are traditionally drunk with milk in cultures where dairy products are consumed. These include Indian masala chai and British tea blends. These teas tend to be very hearty varieties of black tea which can be tasted through the milk, such as Assams, or the East Friesian blend. Milk is thought to neutralise remaining tannins and reduce acidity. The Han Chinese do not usually drink milk with tea but the Manchus do, and the elite of the Qing Dynasty of the Chinese Empire continued to do so. Hong Kong-style milk tea is based on British colonial habits. Tibetans and other Himalayan peoples traditionally drink tea with milk or yak butter and salt. In Eastern European countries (Russia, Poland and Hungary) and in Italy, tea is commonly served with lemon juice. In Poland, tea with milk is called a bawarka ("Bavarian style"), and is often drunk by pregnant and nursing women. In Australia, tea with milk is white tea.

The order of steps in preparing a cup of tea is a much-debated topic, and can vary widely between cultures or even individuals. Some say it is preferable to add the milk before the tea, as the high temperature of freshly brewed tea can denature the proteins found in fresh milk, similar to the change in taste of UHT milk, resulting in an inferior-tasting beverage. Others insist it is better to add the milk after brewing the tea, as black tea is often brewed as close to boiling as possible.

The addition of milk chills the beverage during the crucial brewing phase, if brewing in a cup rather than using a pot, meaning the delicate flavour of a good tea cannot be fully appreciated. By adding the milk afterwards, it is easier to dissolve sugar in the tea and also to ensure the desired amount of milk is added, as the colour of the tea can be observed. Historically, the order of steps was taken as an indication of class: only those wealthy enough to afford good-quality porcelain would be confident of its being able to cope with being exposed to boiling water unadulterated with milk. Higher temperature difference means faster heat transfer so the earlier you add milk the slower the drink cools. A 2007 study published in the European Heart Journal found certain beneficial effects of tea may be lost through the addition of milk.

Many flavourings are added to varieties of tea during processing. Among the best known are Chinese jasmine tea, with jasmine oil or flowers, the spices in Indian masala chai, and Earl Grey tea, which contains oil of bergamot. A great range of modern flavours have been added to these traditional ones. In eastern India, people also drink lemon tea or lemon masala tea. Lemon tea simply contains hot tea with lemon juice and sugar. Masala lemon tea contains hot tea with roasted cumin seed powder, lemon juice, black salt and sugar, which gives it a tangy, spicy taste.

Adding a piece of ginger when brewing tea is a popular habit of Sri Lankans, who also use other types of spices such as cinnamon to sweeten the aroma. Other popular additives to tea by the tea-brewer or drinker include sugar, liquid honey or a solid Honey Drop, agave nectar, fruit jams, and mint. In China, sweetening tea was traditionally regarded as a feminine practice. In colder regions, such as Mongolia, Tibet and Nepal, butter is added to provide necessary calories.

Tibetan butter tea contains rock salt and dre, a butter made from yak milk, which is churned vigorously in a cylindrical vessel closely resembling a butter churn. The same may be said for salt tea, which is popular in the Hindu Kush region of northern Pakistan.

Alcohol, such as whisky or brandy, may also be added to tea. The flavour of the tea can also be altered by pouring it from different heights, resulting in varying degrees of aeration. The art of high-altitude pouring is used principally by people in Northern Africa (e.g. Morocco, Algeria, Mauritania, Libya and Western Sahara), but also in West Africa (e.g. Guinea, Mali, Senegal) and can positively alter the flavour of the tea, but it is more likely a technique to cool the beverage destined to be consumed immediately. In certain cultures, the tea is given different names depending on the height from which it is poured. In Mali, gunpowder tea is served in series of three, starting with the highest oxidisation or strongest, unsweetened tea (cooked from fresh leaves), locally referred to as "strong like death", followed by a second serving, where the same tea leaves are boiled again with some sugar added ("pleasant as life"), and a third one, where the same tea leaves are boiled for the third time with yet more sugar added ("sweet as love"). Green tea is the central ingredient of a distinctly Malian custom, the "Grin", an informal social gathering that cuts across social and economic lines, starting in front of family compound gates in the afternoons and extending late into the night, and is widely popular in Bamako and other large urban areas.

In Southeast Asia, particularly in Singapore and Malaysia, the practice of pouring tea from a height has been refined further using black tea to which condensed milk is added, poured from a height from one cup to another several times in alternating fashion and in quick succession, to create a tea with entrapped air bubbles creating a frothy "head" in the cup. This beverage, teh tarik, literally, "pulled tea" (which has its origin as a hot Indian tea beverage), has a creamier taste than flat milk tea and is extremely popular in the region.

Tea pouring in Malaysia has been further developed into an art form in which a dance is done by people pouring tea from one container to another, which in any case takes skill and precision. The participants, each holding two containers, one full of tea, pour it from one to another. They stand in lines and squares and pour the tea into each other's pots. The dance must be choreographed to allow anyone who has both pots full to empty them and refill those of whoever has no tea at any one point.

ECONOMICS BENEFIT OF TEA

Tea is the most popular manufactured drink consumed in the world, equaling all others – including coffee, chocolate, soft drinks, and alcohol – combined. Most tea consumed outside East Asia is produced on large plantations in the hilly regions of India and Sri Lanka, and is destined to be sold to large businesses. Opposite this large-scale industrial production are many small "gardens," sometimes minuscule plantations, that produce highly sought-after teas prized by gourmets. These teas are both rare and expensive, and can be compared to some of the most expensive wines in this respect.

India is the world's largest tea-drinking nation, although the per capita consumption of tea remains a modest 750 grams per person every year. Turkey, with 2.5 kg of tea consumed per person per year, is the world's greatest per capita consumer.

PRODUCTION OF TEA

In 2003, world tea production was 3.21 million tonnes annually. In 2010, world tea production reached over 4.52 million tonnes after having increased by 5.7% between 2009 and 2010. Production rose by 3.1% between 2010 and 2011. The largest producers of tea are the People's Republic of China, India,

Kenya, Sri Lanka, and Turkey. The following table shows the amount of tea production (in tonnes) by leading countries in recent years. Data are generated by the Food and Agriculture Organization of the United Nations as of February 2012.

Rank	Country	2008	2009	2010	2011
1	China	1,257,600	1,359,000	1,450,000	1,623,000
2	India	987,000	972,700	991,180	1,063,500
3	Kenya	345,800	314,100	399,000	377,912
4	Sri Lanka	318,700	290,000	282,300	327,500
5	Turkey	198,046	198,601	235,000	221,600
6	Vietnam	173,500	185,700	198,466	206,600
7	Iran	165,717	165,717	165,717	162,517
8	Indonesia	150,851	146,440	150,000	142,400
9	Argentina	80,142	71,715	88,574	96,572
10	Japan	96,500	86,000	85,000	82,100
Total	World	4,211,397	4,242,280	4,518,060	4,321,011

Labor and Consumer Safety Problems

Multiple recent reports have found that most Chinese and Indian teas contain residues of banned toxic pesticides. Tea production in Kenya, Malawi, Rwanda, Tanzania, and Uganda has been reported to make use of child labor according to the U.S. Department of Labor's List of Goods Produced by Child Labor or Forced Labor (a report on the worst forms of child labor).

Certification

Workers who pick and pack tea on plantations in developing countries can face harsh working conditions and may earn below the living wage. A number of bodies independently certify the production of tea. Tea from certified estates can be sold with a certification label on the pack. The most important certification schemes are Rainforest Alliance, Fairtrade, UTZ Certified, and Organic, which also certify other crops such as coffee, cocoa and fruit. Rainforest Alliance certified tea is sold by Unilever brands Lipton and PG Tips in Western Europe, Australia and the US. Fairtrade certified tea is sold by a large number of suppliers around the world. UTZ Certified announced a partnership in 2008 with Sara Lee brand Pickwick tea. Production of organic tea has risen since its introduction in 1990 at Rembeng, Kondoli Tea Estate, Assam. 6,000 tons of organic tea were sold in 1999. About 75% of organic tea production is sold in France, Germany, Japan, the United Kingdom, and the United States.

TRADE

According to the FAO in 2007, the largest importer of tea, by weight, was the Russian Federation, followed by the United Kingdom, Pakistan, and the United States. Kenya, China, India and Sri Lanka were the largest exporters of tea in 2007 (with exports of: 374229, 292199, 193459 and 190203 tonnes respectively). The largest exporter of black tea is Kenya, largest producer (and consumer) India.

PACKAGING OF TEA

Tea bags

In 1907, American tea merchant Thomas Sullivan began distributing samples of his tea in small bags of Chinese silk with a drawstring. Consumers noticed they could simply leave the tea in the bag and reuse it with fresh tea. However, the potential of this distribution/packaging method would not be fully realised until later on. During World War II, tea was rationed in the United Kingdom. In 1953 (after rationing in the UK ended), Tetley launched the tea bag to the UK and it was an immediate success.

The "pyramid tea bag" (or sachet) introduced by Lipton and PG Tips/ Scottish Blend in 1996, attempts to address one of the connoisseurs' arguments against paper tea bags by way of its three-dimensional tetrahedron shape, which allows more room for tea leaves to expand while steeping. However, some types of pyramid tea bags have been criticised as being environmentally unfriendly, since their synthetic material is not as biodegradable as loose tea leaves and paper tea bags.

LOOSE TEA

The tea leaves are packaged loosely in a canister, paper bag, or other container such as a tea chest. Some whole teas, such as rolled gunpowder tea leaves, which resist crumbling, are sometimes vacuum packed for freshness in aluminised packaging for storage and retail. The loose tea must be individually measured for use, allowing for flexibility and flavor control at the expense of convenience. Strainers, tea balls, tea presses, filtered teapots, and infusion bags prevent loose leaves from floating in the tea and over-brewing. A traditional method uses a three-piece lidded teacup called a gaiwan, the lid of which is tilted to decant the tea into a different cup for consumption.

COMPRESSED TEA

Compressed tea (such as Pu-erh) is produced for convenience in transport, storage, and ageing. It can usually be stored longer without spoilage than loose leaf tea. Compressed tea is prepared by loosening leaves from the cake using a small knife, and steeping the extracted pieces in water. During the Tang dynasty, as described by Lu Yu, compressed tea was ground into a powder, combined with hot water, and ladled into bowls, resulting in a "frothy" mixture. In the Song dynasty, the tea powder would instead be whisked with hot water in the bowl. Although no longer practiced in China today, the whisking method of preparing powdered tea was transmitted to Japan by Zen Buddhist monks, and is still used to prepare matcha in the Japanese tea ceremony. Compressed tea was the most popular form of tea in China during the Tang dynasty. By the beginning of the Ming dynasty, it had been displaced by loose leaf tea. It remains popular, however, in the Himalayan countries and Mongolian steppes. In Mongolia, tea bricks were ubiquitous enough to be used as a form of currency. Among Himalayan peoples, compressed tea is consumed by combining it with yak butter and salt to produce butter tea.

INSTANT TEA

"Instant tea", both hot and cold, is an alternative to the brewed products. Similar to freeze-dried instant coffee, but not requiring boiling water, instant tea was developed in the 1930s. Nestlé introduced the first commercial product in 1946, while Redi-Tea debuted instant iced tea in 1953.

Delicacy of flavour is sacrificed for convenience. Additives such as chai, vanilla, honey or fruit, are popular, as is powdered milk. During the Second World War British and Canadian soldiers were issued an instant tea known as 'Compo' in their Composite Ration Packs. These blocks of instant tea, powdered milk, and sugar were not always well received. As Royal Canadian Artillery Gunner, George C Blackburn observed:

- But, unquestionably, the feature of Compo rations destined to be remembered beyond all others is Compo tea...Directions say to "sprinkle powder on heated water and bring to the boil, stirring well, three heaped teaspoons to one pint of water."
- Every possible variation in the preparation of this tea was tried, but...it always ended up the same way. While still too hot to drink, it is a good-looking cup of strong tea. Even when it becomes just cool enough to be sipped gingerly, it is still a good-tasting cup of tea, if you like your tea strong and sweet. But let it cool enough to be quaffed and enjoyed, and your lips will be coated with a sticky scum that forms across the surface, which if left undisturbed will become a leathery membrane that can be wound around your finger and flipped away...

BOTTLED AND CANNED TEA

Canned tea is sold prepared and ready to drink. It was introduced in 1981 in Japan. In 1983, Swiss-based Bischofszell Food Ltd., was the first company to bottle ice tea on an industrial scale.

STORAGE OF TEA

Storage conditions and type determine the shelf life of tea. Black tea's is greater than green's. Some, such as flower teas, may last only a month or so. Others, such as pu-erh, improve with age. To remain fresh and prevent mold, tea needs to be stored away from heat, light, air, and moisture. Tea must be kept at room temperature in an air-tight container. Black tea in a bag within a sealed opaque canister may keep for two years. Green tea deteriorates more rapidly, usually in less than a year. Tightly rolled gunpowder tea leaves keep longer than the more open-leafed Chun Mee tea. Storage life for all teas can be extended by using desiccant or oxygen-absorbing packets, vacuum sealing, or refrigeration in air-tight containers (except green tea, where discrete use of refrigeration or freezing is recommended and temperature variation kept to a minimum).

GREEN PLANT TO FINISHED PRODUCT AND TEA

In India, black is manufactured and used. Green leaf tea is also manufactured which is mainly exported. This document is mainly eccentric to black tea manufacturing.

NURSERY

A good plant can be developed and grown only under a regime of good agromanagement in the nursery, till it is transplanted to the field. Therefore, a nursery must not only to provide the most optimum conditions for growth of the plants but the chosen sites should also be convenient for the area to be planted to minimise any damage to the plants during transit, and indeed minimise transport cost. It should be reasonably leveled with no risk of waterlogging during heaviest of rains.

The soil should be chemically suitable with a good structure. A deep top soil normally ensures a good depth of fertile soil in the bed. A useful nursery requires all weather access, shade, drained beds and water. Sandy loam soil is the best for raising a nursery on orthodox beds though virgin soil will provide the best medium. Extreme soils at best are avoided. On heavy soils seedling growth is rather slow, though by adding sand or cattle manure, plant growth in these soils could be hastened within limits.

Dark coloured top soil is normally rich in fertility but soil very poor in organic matter may do with a little dry cow dung or a fine compost blended in. A useful practice is to sieve the soil through No. 3 mesh to remove undesirable elements like old bits of roots. It is equally important that the soil is not drying out, turning muddy or is compressed as these might affect the growth of the plants. To facilitate growth, 500 gm of super-phosphate per cubic metre soil would be advantageous; other fertilisers may be incorporated if analysis shows a specific need.

In any case, no soil should be used which has a high proportion of eelworms. Current trend is to grow tube nurseries rather than bed nurseries as the management of plants is much easier when they are grown in tubes. Although size of the tube is a matter of convenience, an optimum size should ensure good growth. Six to eight weeks before their use the sleeves are filled ideally with good textured medium-loam soil which ensures that water does not accumulate within the sleeves. If more than half of the soil is fine silt the sleeve soil will have to have sand. The pH value of the soil should remain between 4.5 and 5.5. Extreme values outside this range should be adjusted with lime or aluminium sulphate. For nursery plants, a low level of shade is essential but they should give a clearance of 25 cm above sleeve tops. Shade used in Assam and the Dooars consists of 1.30 x 1.30 m bamboo basket laid across the bed fences. Winter planting of nurseries can be started without shade, but a light mulch is needed to prevent drying out of the bed. Eventually, a shade has to be provided not only to prevent desiccation of the soil but also to provide appropriate growth stimuli to the nursery plants as well. But more than shade per se, a proper regulation of shade is essential for the growth of the nursery plants.

In general, as the nursery plants get themselves established shade, can be selectively reduced, and is finally removed when it interferes with the growth sequence of the plants. The latter is done when the nursery plants have produced four to five good leaves but prior to that the process has to be gradual to allow the growing nursery plants in getting accustomed and adjusted to intense sunlight. Shade is removed differentially first when there is a two leaf growth. This is done first on a cloudy day and finally the shade is removed when the plants have four to five leaves growth. At higher elevations, 1,400 m and above, a polythene tent supported on bamboo strips and hoops 45 m above the bed would be useful. Application of fertilisers will increase the rate of growth. When soil in the tube or bed is fertile, nitrogen alone will have a significant effect.

Phosphate is also effective is most situations. A soluble NPK mixture for foliar application (10 kg urea plus 20 km ammonium phosphate plus 8 kg murate of potash) is applied from unfolding of two leaves till plants have eight leaves. Weeds can be a problem when plants are small. At the very incipient stage weeds can be removed physically, though a some herbicide will effectively control the weeds for a longer duration. Applications of insecticides, acaricides and fungicides will depend on specific pest and disease situations. Disbudding or tipping is useful for encouraging laterals, but this should be done cautiously to avoid any damage to growing plants. This is best done when plants are 30 to 35 cm high with step up.

FIELD PLANTING

From all accounts planting is a most crucial operation as it basically determines the development and productive level of tea throughout its economic life. Wrong planting of good planting materials is doubly unproductive as investment is lost both on account of producing the plants and in the failure to put them up for productivity. Therefore, care, planning and refinement of techniques are essential for long-term benefit. The very fundamental to planting, apart from good planting materials, is that the planting hole should be of adequate size. The size of the hole partly depends on the consistency of the soil but a diameter of 45 cm and a depth of 75 cm is ideal; the hole could be dug in three phases - the first phase is dug to 15 cm involving cutting of top soil with a hoe and placing it separately; the second phase involves digging up to 30 cm which involves subsoil layer -the excavated portion is also maintained separately. In the third phase the balance 30 x 15 cm across is loosened with posthole auger or deep forking. The size of hole also depends on the consistency of the soil but in general a standard size of 45 x 75 cm ought to be maintained.

When the holes are being dug the crumb structure of the soil is often damaged, but alternate wetting and drying of the puddle system corrects this. However, if the excavated holes are exposed for a long time without refilling, the excavated soil may be lost due to erosion. Therefore, the holes should be dug with sufficient time in hand to allow the soil to settle down. For closer spaced planting and heavy soils, an alternative to holing is trench planting. Trenches 30 cm wide and 45 cm deep are dug along rows either manually or mechanically. Trenches provide a lower volume of soil for root expansion. The holes are refilled two or three weeks before planting with top soils mixed with dry old cattle manure (4 kg) or 5 kg well-composted organic matter mixed with 30 to 40 gm of superphosphate. Since the initial growth of the plants is dependent on their food reserve rather than mineral nutrition of the soil, organic manures would be valuable. However, care must be taken to avoid having a layer of compost at the bottom of the pits as it may cause localised water logging and ensure that high concentration of mineralising organic matter does not produce a high level of pH.

BRINGING UP OF YOUNG TEA PLANTS AND BUSH FORMATION

When a young plant is established in the field it is essential to train or modify it to form a low bush of good spread and of convenient height for plucking. The essence therefore is to establish a good frame and ensure a good coverage of ground within the shortest possible time. This is achievable either by de-centering and pruning, or by bending or pegging, or by a combination of de-centering and bending.

The decentering and pruning combination involves cutting off of the main stem at a height between 7 cm and 20 cm - the few laterals remaining on the stem are cut at 35 cm above ground level. This excision stimulates the plants to throw out more laterals from the lower portion of the stem, thus increasing the number of pruning sticks. The plants are next pruned within two years after the first prune for de-centering, and this is done about 5 cm above the last prune mark. This part of the operation, that is, decentering, can be done as soon as the plants start producing new growth after transplanting, and in more vigorous plants this can be done even earlier than in weaker plants.

In hilly areas, where plant growth rate is generally slow de-centering is done in spring or autumn after planting. A drawback of de-centering process is that it might cause high mortality unless the full vigour of the plants is assessed. The argument in favour of bending and pegging is that they do not cause heavy mortality during the operation, provide quick ground cover, apart from producing a substantial yield by the second year. For vigorous plants, pegging can be done two to three weeks after planting. Shoots that are 30 to 60 cm high are bent at angles 30° to 45° to the ground, and are then pegged down to the ground. All lateral shoots produced subsequent to the bending are also bent and pegged. But bending could be difficult with overgrown plants and is risky in un-shaded areas because of scorching risks. Moreover, in contour planting, it is always desirable to have the 'bends' along the contour, preferably turned towards west or south. The pegged main steam and the laterals are first pruned 30 to 40 cm from the ground within two to three years, and are then raised by 5 cm in each successive prune till the permanent frame height is reached.

Three variants of pegging are practiced, at least in north-east India. First, bending with single pegs, which is a laborious process involving repeated pegging using single pegs either of bamboo, wood or bamboo. It is always difficult to keep the pegs in position as often they are removed by the growing pressure of the branches. In the ring pegging method after the initial bending of the plant by a single peg, or following a de-center, the laterals that emerge are allowed to spread out by putting a ring of bamboo strips 25 to 30 cm diameter in the middle. This ring is often kept in position by three to four pegs but the process is rather cumbersome. In 'rail' pegging method, the laterals and re-growth are repeatedly bent pushing them beneath two rails of bamboo strips. Though easy to operate, the main drawbacks here are the difficulties in maintaining the desired angle. Yet a third approach towards frame formation is to de-centre first and then go for pegging. This essentially involves de-centering of the main stem with the laterals being bent to the sides and pegged. Subsequently, pruning is done at 35 to 50 cm from the ground and then the frame is raised by 5 cm at each successive prune till a permanent frame is established.

One advantage of this system is that it is easier to train the branches as less number of pegs are needed. On account of decentering in the first year there could be an initial yield loss of about 18 percent, but the subsequent rise in crop well compensates the earlier losses. As pointed out earlier, the main objective in rearing young tea is to divert apical dominance of the main stem and force the plants to develop well-spread lateral branches. To achieve this twain objective a "low tipping method" was advocated. This technique involves total avoidance of pruning in the first four years of planting, centering out within 10 weeks after planting and initial plucking at a low level of 40 to 50 cm from the ground. The first pruning is considered only when the plants are above 85 cm high at the end of the fourth or fifth year with pruning height varying between

55 and 65 cm from the ground. This technique resulted in high yield in the early years but it failed to give a food frame so essential for long-term productivity of tea.

Therefore, by giving a prune, rather than totally avoiding it in the early years, a good frame could be established. With the introduction of bending or pegging, good ground cover, ideal frame and early crop are possible to achieve. Whether pegging would be ideal with high plant populations, irrespective of the growth habits of different cultivars, is not clear. Three other systems used in varying degrees are centering out, lung-pruning and de-budding. Centering out involves cutting out of the main stem at a height of 15 to 20 cm to force lateral growth from below the cut.

This operation can be done six to eight weeks after planting or when the plants have produced a few flushes or leaves indicating that they have established themselves. Since the operation is rather drastic, it is most important that starch reserve of the plants is at its optimum. If, per chance, only a few leaves have developed, then it is better to avoid this system and go for either lung-pruning or de-budding.

Lung-pruning or thumb-pruning as it is popularly known, is a mechanical operation. When plants are between 15 and 20 cm high, they are held between the thumb, and the stem is half broken, making sure that one side is kept intact for movement of food and nutrient. The upper part of the stem is then bent towards the ground and the broken portion is removed only after the plant produces a flush of new growth. Unlike centering out operation, lung-pruning can be done even if starch content is not at its optimum.

De-budding, as the name implies, consists of removal of the apical bud and all buds in leaf axils, 15 to 25 cm above the ground. This ensures that no laterals grow from the upper part of the plant and after laterals have grown from the lower portion, the stem is removed at 15 to 20 cm. De-budding is an extremely delicate operation and can even be done in the nursery in situations where growth of laterals is a problem.

It is difficult to generalise on the relative merits and demerits of different systems of raising young tea because the principal source of variation, that is, the growth characteristics of different cultivars have not been considered in most studies, despite marginal evidence that some tea varieties behave alike under a particular system of bush management. This needs thorough elaboration using cultivars of different genotypic origin. Nonetheless, by adopting one or other system, the growth rate of young tea can be enhanced ensuring early yield.

On relative efficacies of the four principal methods — low tipping and step up plucking; low tipping with frame forming prune and step up plucking; pegging and step up plucking; and pegging after de-centre with step up plucking, the general views are:

- Low tipping produces good early crop but frame formation is poor, and frame formation prune 18 months after planting produce good frame with high yields from fifth year onwards.
- Pegging has an advantage in terms of crop increase only in the two initial years. Thereafter, this advantage is lost and the system's overall performance equals the combination of low tipping and frame formation pruning.
- De-centering before pegging causes an immediate crop loss but this is wellcompensated later. This apart, pegging by itself may cause serious sun scorch which in some cases may even remain permanently on the frame.

NECESSITY OF PRUNING

Perhaps from a productive point of view no field operation is as important as pruning is, as it basically helps in maintaining the plant as low bush in a phase of continuous vegetative growth. Pruning therefore is considered as "at best a necessary evil" as it both stimulates and controls growth. The other principal objectives of pruning are :

- Shaping the tree to make the best use of space between trees.
- Reestablishing the initial vigour of the shoot system to stimulate growth, and control undue rise in bush frame and plucking table, that is, control growth.
- Removing dead, diseased and overage wood, thus help rejuvenate bushes that have crossed the period of maximum productivity.
- Providing an environment within the plant which is conductive to maximum crop production but minimises the spread of pests and diseases.
- Having the crop formed where it can be harvested easily and cheaply.
- Renewing the actively growing branches and maintaining sufficient volume of maintenance foliage to meet physiological needs of the plant.
- Maintaining quality in made tea.

It is generally accepted that bushes are best pruned during December to January and must not be rested earlier than October. Resting during a phase of slow growth is essential to create a favourable carbohydrate balance, but this principle does not hold good for all countries, that is, the timing of pruning varies depending on local growing conditions. The correct timing for pruning is decided basically from an assessment of the growth rate of the plants. It should be at a point when growth has sufficiently slowed down to create a favourable carbohydrate balance, so that, a layer of foliage could grow and get established well before the hot spell. The other factors that influence pruning time are prevention from potential dangers of sun-scorch, drought or attacks from pests and diseases, apart from the more mundane considerations like the re-growth time desired and the crop planned. In summarising, pruning is a means for reducing the number of stalks on the bushes, so that, an optimum quality of wood is left for giving the best and most vigorous shoots the following year. The amount of wood left will again vary according to the growth characteristics and vigour of the cultivars but leaving 3 to 4 cm of wood may be useful under certain situations.

CUTTING TYPES

Pruning

Collar Pruning

Collar pruning is the severest form of pruning, requiring as it does, cutting the bole of the bush near or at the ground level. In some cases, the cut may even be given slightly below the ground level to ensure protection of the exposed tissues against die-back. Recovery from this severe form of pruning is greatly influenced by carbohydrate reserve. By and large, collar pruning is suitable for tea that is characterised by a well marked dormancy period. The main objective is to provide a new fame to the plant by removing old and unproductive branches affected by diseases and pests, particularly in situations where uprooting of unproductive bushes is not possible. Tea recovering from collar pruning also needs special care and management. In heavy pruning, which is a variant of collar pruning, the bole is cut at 15 to 45 cm above the ground for frame renewal. But this kind of hard pruning can only be done after serious consideration of plant health and carbohydrate reserve. A debilitated plant exposed to extreme forms of pruning may cause further depletion of food reserve without producing any compensatory maintenance foliage, and in the process the frame formation would suffer.

Medium Pruning

The main objective of medium pruning is to reduce height of the bushes, so that, they do not exceed 80 cm. Continuous picking of shoots on the top of a bush gradually increases the height of the bushes, apart from causing congestion on the top with weak and twiggy branches. This drain in the physiological resources which otherwise could have been utilised for increasing productivity. Therefore, to bring back the plants to a productive level, stimulate new wood and renew maintenance foliage, medium pruning becomes a useful means. Though relatively less severe than collar or heavy pruning, medium pruning is generally not done below 50 cm, and certainly not lower than what is necessary to remove the knots and congested top. As with collar pruning, a major determinant for medium pruning is to decide that the plants are ready to take this drastic treatment, that is, whether they have adequate carbohydrate reserves, and whether chances of scorching are minimal. In north-east India, this operation is ideally carried out during end December to end January, but before this operation is carried out, plants must be rested for six to eight weeks.

Top or Light Pruning

This is the standard method of recurring pruning in Assam and the Dooars. The main objective of this pruning is to renew the leaf bearing branches, apart from cleaning out the bush. Normally top pruning is done 2 to 3 cm above the last prune; in effect this system of pruning leaves a considerable amount of foliage on the plants though much of it defoliates naturally; but under certain conditions of growing this foliage may even remain infact and carry on normal physiological functions. The technique of light pruning therefore essentially involves a uniform cut-across followed by a cleaning out operation to remove thin unproductive branches and dormant shoots. Sometimes only a cut-across, that is, removal of top part of light pruned tea is practiced to invigorate growth without undue removal of foliage.

Lung Pruning

The essence of lung pruning is to prune selectively, so that, all leaves below the level plucking are left and the peripheral branches are not cut. These left out branches have turned lungs though in no way are they connected with respiration per se. This system of pruning ensures that sufficient amount of actively synthesising foliage is left on the bush so that a carbohydrate balance is maintained to enable optimal growth of the plants. The use of this system of pruning is rather restricted, and leaving a few 'breathers' at the time of medium pruning have generally been beneficial for stimulating growth.

Skiffing

Skiffing is the lightest form of pruning involving as it does removal of certain amount of growth above the previous pruning level. It essentially is the levelling of the foliage either by giving a cut in the green stem in the lighter forms of skiffing, or, into the red wood in harder forms of skiffing. The object of skiffing is not only to have an early start of crop production but also to achieve more of the high quality first flush and total crop as well. Skiffing helps in thickening the pruning wood and improving the general health of the bushes.

Tipping

Immediately after pruning, the plants enter into a phase of unrestricted growth and if this unrestricted growth is allowed, most of the shoots would mature as branches. To prevent this, the new growth has to be controlled, so that a uniform plucking table is formed. Tipping therefore is an operation aimed at forming a level and a flat plucking surface, and filling it by quick production of secondary branches.

PLUCKING GREEN TEA

Plucking or picking denotes harvesting of tea crop. Plucking of young shoots, the two leaves and buds, is an complex art consuming about 15 percent of the total cost of production of tea and much of the quantity and quality of harvested crop depends on the standard of plucking. It involves removal of young growing shoots comprising the apical bud, the internodes and two or three leaves below it, which together constitute the crop. The tender shoots removed are the sinks as they cannot support their normal growth. They are harvested at regular intervals to stimulate successions of a new crop of shoots. The growing shoots in turn are dependent on the assimilates from the mature foliage where they are produced, that is, the source. Consequently, an ideal plucking system must take cognizance of the intricate relationship between the sink and the source. In practical terms, the amount of maintenance foliage to be retained is of prime consideration in determining the system of plucking to be adopted as it is the maintenance foliage that provides the assimilates for the production of shoots. The growing bud exerts maximum pulling force for metabolites from the maintenance foliage; the relative sink capacity of a shoot of one leaf and a bud and two leaves and a bud being 75 and 35 percent, respectively. Therefore, indiscriminate removal of large buds and one leaf and a bud would adversely affect the health of bushes. The initial growth of leaves following pruning occurs at the expense of carbohydrates formed during the previous growth and stored in roots and older wood. These new shoots develop from the dormant buds on the stubs that are left on the plant during pruning. As explained earlier, these crops of new shoots are the primaries and the two covers on the dormant buds are known as the scale leaves or cataphylls or janams. The emerging leaves are arranged in spirals around the stem in such a manner that the first five leaves form a series which encircle the stem twice, while the sixth leaf appears above the first of the previous series.

PLUCKING STANDARD

The standard of plucking is unequivocally important as more than anything else, it determines the quality potential of the processed leaf. It is generally accepted that fine plucking makes the best quality tea as this form of plucking consists picking of only tow leaves and a bud and plucking of these flushes is regarded as fine plucking. In a broader sense, fineness of plucking may also involve harvesting all one and a bud, two and a bud and single banjhis over the janam, apart from the fact that fine plucking also calls for plucking at short intervals usually at seven days intervals. If still shorter rounds of plucking are practiced to get leaves with attributes for high quality, then it may cause severe depletion of 'sinks' on plucking surface as the shoot will not have opportunities for optimal growth so necessary before plucking.

Consequently, over a period of time total yield may decline but quality of tea will certainly improve. At the other extreme of fine plucking are the medium, coarse or very coarse plucking. These three systems of plucking greatly increase the quantity of harvested crop, but only at the cost of quality. Medium plucking, for example, involves harvesting of large shoots with two leaves and a bud, three leaves and a bud plus the single banjhis or dormant leaves. Coarse plucking involves removal of all shoots with two leaves and a bud, three leaves and a bud, four leaves and a bud and double banjhis. Very coarse plucking removes all shoots with three leaves and a bud and even shoots with four leaves and a bud. Coarse plucking can be done at 8 to 10 or even 12 days intervals, and because of the increase in the size of shoots, the weights of harvest also increase proportionately with the length of the shoots harvested.

Indeed, a linear relationship, exists between the size of shoots with different combinations of leaves and a bud and their weights. In terms of 'sink' capacity of pluckable shoots, shoots with one leaf and a bud have the highest sink capacity - therefore, their indiscriminate removal along with larger buds may adversely affect the physiological balance causing a decline in the production of subsequent flushes. From this point of view at least standard plucking would appear to be ideal.

PLUCKING INTERVAL

For efficient harvesting of shoots plucking interval is most crucial. The interval is ideally determined by the rate of a leaf unfolding. In north-east India, the general practise is to harvest shoots at 7 days intervals on the assumption that by this time the levels of various biochemical constituents necessary to make tea of quality would reach their optima. Any increase or decrease in plucking interval could significantly influence the dynamic metabolic system, even leading to alterations in leaf physiology. Therefore, plucking rounds at shorter intervals with fine plucking can reasonably assure a better quality of made tea than teas plucked at longer intervals. Thus, 5 days plucking rounds can give good quality of made tea because of a lower content of ash and fibre, though black teas made from 7 days pluckings would have a balance of VFC , ash, soluble solids and caffeine and therefore would have an optimum quality. Teas manufactured from 9 to 11 days pluckings are generally of low quality possibly because of a decrease in soluble solids, with a significant increase in crude fibre.

It is equally important that theaflavin also vary with the maturity of leaves at 5 to 11 days plucking intervals. Apart from quality, plucking interval also affects overall quantum of harvestable shoots. Generally, as the plucking interval or length is increased, there is a corresponding increase in yield of both light pruned and unpruned tea, but quality deteriorates. This happens because longer rounds increase the quantum of coarser leaf and banjhi leaves, and both have a negative effect on quality. This drawback notwithstanding, plucking interval is often extended to 8 to 10 days in early spring or later autumn, or shortened to 5 to 6 days, depending on the growth of shoots, or on purely mundane consideration like the requirements of quantity or quality of tea at a point of time. What therefore is an ideal plucking round is difficult to generalise, because as stated earlier, very many interacting phenomena are involved here. But where standard plucking is the practice, the shortest possible round could be 5 days during the monsoon , the longest being 7 days. With fine plucking the corresponding period should be 6 and 9 days respectively.

BLACK TEA PROCESSING

OVERALL STEPS

Black tea manufacturing technology involves disruption of the cellular integrity of tea shoots; in doing so, the mixing of substrates, polyphenols and the enzymes, polyphenols oxidase is facilitated. This results in the initiation of a series of biochemical and chemical reactions with the uptake of atmospheric oxygen and formation of pigmented hot water soluble polyphenolic compounds, characteristic of black tea. Black tea processing consists of the following unit operations:

- Withering
- Rolling/disruption
- Oxidation/Fermentation (biochemical reactions in the presence of oxygen)
- Drying
- Sorting (fibre removal; grading based on size) Of these, chemical changes occur primarily during withering, fermentation and drying.

Withering

Withering is carried out to prepare the green tea shoots bio-chemically and physically for subsequent manufacturing operations. Withering essentially consists of storage of green shoots for about 12 to 20 hours with the partial removal of moisture from the leaf. Withering is accompanied by certain chemical changes which may affect fermentation and thereby, the quality of the final product. Chemical wither is essentially a timetemperature dependent process. Some of these changes are dependent on moisture loss while others are independent. The major physical change during withering is the increase in cell membrane permeability and consequently, the leaf becomes flaccid. The increased permeability of the membrane has a great effect on the mixing of substrates and enzymes during fermentation. Withering is also accompanied by the activation of oxidative, hydrolytic and proteolytic enzymes, causing significant changes in the chemical composition of tea shoot, as a result the levels of soluble protein, free amino acids and simple sugars increase in the withered leaf. Amino acids increase due to the breakdown of proteins by the enzyme pepsidase. Amino acids are primarily the precursors of the aromatic compounds.

Caffeine content increases with withering. The increase in caffeine content is greater if the withering is accompanied by loss of moisture. Caffeine forms a complex with theaflavins thearubigins and "Creams" down from the liquor; this creaming down is a desirable cup character for black tea. The Chemical wither also increases the level of organic acids and improves polyphenol oxidase (PPO) activity.

The increase in organic acids will have an obvious effect of reducing pH and TF formation is increased as pH tends towards 5.0. The effect of increased PPO activity is beneficial for more TF formation provided oxygen availability is not limited during catechin oxidation. Volatile compounds essentially impart aroma to the finished product. Volatile compounds present in the tea are classified into two groups : Group-I compounds, which are deleterious to the tea quality and Group-II compounds, which impart sweet flowery aroma to tea and its presence is highly desirable. The ratio of Group-II compounds to Group-I compounds is known as the flavour index and it is a semi-quantitative method of describing the aroma quality of tea. Tea with higher values of flavour index normally have better aroma quality. Withering improves flavour index and thereby, sweet aroma.

Leaf Disruption

This stage in processing is described in a number of ways, e.g. rolling, cutting, crushing, tearing, but the basic requirements are size reduction with a degree of cell disruption to allow the exposure of the new surfaces to air in the subsequent fermentation stage. The LTP machine consists of a central rotor with a number of knives and beaters, where the leaf is bruised and cut by the rotating knives and beaters. The leaf next enter into the rolling section. The rolling is done either by orthodox rollers or by a combination of orthodox rolling and the CTC machines. The surface of each roller is made up of a number of segments made of stainless steel. The withered leaf passes through a battery of three to five such CTC machines which cause the leaf to get crushed, torn and curled.

Fermentation

Tea fermentation is essentially the oxidation of simple substrates into complex characteristic substances by endogenous enzymes present in the tea leaf. This distinguishes tea manufacture from other food processes in which exogenous fermentation inducing agents are added to the raw material. Fermentation is not confined only to the period during which the leaf lies in the fermenting drums, racks or floors. It commences from the time the cells are bruised and extends until the enzymes are deactivated in the drier. As a first step during fermentation, the catechins are oxidized to highly reactive transient orthoquinones by polyphenol oxidase.

The quinones in turn dimerize to produce theaflavins (TF), which are orange red substances that contribute significantly to astringency, briskness, brightness and the colour of tea beverage. Theaflavins comprise 0.3 to 2.0 percent of the dry weight of black tea. Further transformations of dicatechins and theaflavins yield compounds which are known as thearubigins, comprising about 9 to 19 percent of black tea, are red brown in colour and contribute to colour, strength and mouth feel of liquor. While theaflavin content of tea increases during fermentation and starts declining after reaching a peak thearubigin content continually increases throughout the fermentation. In actual practices, the completion of fermentation is judged by the change in colour and the pleasant aroma that develops.

However, chemically optimum fermentation may be assessed by monitoring the profile of TF content and taking the time required for the production of maximum. TF Proper balance between theaflavins and thearubigins is also essential for a good cup of tea. Like any other enzymic reaction, the rate of fermentation primarily depends upon the concentration and relative proportion of individual substrates, availability of oxygen, the activity of enzyme, pH and the temperature.

Since it is not a homogeneous system, the degree of contact between the enzyme and substrate will be important. In addition, mass transfer of air to the reaction side is crucial in controlling the rate and product profile of fermentation.

During fermentation, the PPO activity is known to decrease due to the formation of insoluble complexes of polyphenol oxidation products with the enzyme protein by feed back action. This precipitation process increases as fermentation continues and the temperature of the fermentation leaf increases. Both reaction rate and the ultimate concentration of products are affected by the concentration of oxygen. Theaflavin production will be inhibited if the concentration of the oxygen in the air falls below normal, which can occur in dead spots' in a forced-air fermentation system. By increasing the oxygen concentration of the air in a fermentation unit, it is possible to achieve higher theaflavin levels.

Temperature is one of the most important factors influencing the complex series of enzymic and chemical reactions during fermentation. Temperature not only affects the rate of fermentation but also the ultimate level of theaflavin in tea.

As a rule of thumb for altering the fermentation times to suit ambient temperature conditions, the rate of fermentation at any temperature will be taken as 1.7 to 1.8 times as fast as that a 10°C cooler. Though, increase in temperature accelerates the fermentation rate, maximum theaflavin occurs, when fermented at low temperature for a long time. This happens because as the temperature increases, the break-downs reactions become more rapid than the theaflavin formation reaction, and ultimately less theaflavin is produced. Another advantage of low temperature fermentation is that the peak of TF production is maintained for a longer period. Even a delay in firing results in a detectable difference in quality.

Drying

The drying of fermented tea has three major objectives; to terminate the biochemical functions by heat denaturation of the enzyme; to reduce the moisture to increase the shelf stability of black tea and finally, to enhance chemical reactions responsible for black tea character and flavour. The polyphenol oxidase enzymes which convert catechins into theaflavin are not inactivated as soon as the leaf enters the drier. In fact the fermentation process is actually accelerated and continues at a faster rate till the temperature of leaf reaches 55°C at which the enzyme is completely inactivated. Hence, any unoxidised catechin will continue to be converted into theaflavin until oxidase enzyme has been inactivated.

It has been reported that 10 to 15 percent of theaflavin content in black tea is formed during the first 10 minutes of drying. A noticeable effect of drying is the change in colour of dried leaf brought about by the transformation of chlorophyll into pheophytin, which imparts the desired black colour. During drying, the reduction in stringency of fermented tea occurs due to the combination of polyphenols with tea leaf proteins at the elevated drying temperature. Drying causes an overall reduction in the quantity of volatile compounds, although certain aromatic compounds continue to be formed. The fermented Tea leaf is dried by exposing to a draft of hot air; the process takes about 30 minutes. Some of the flavour characteristic of the final product are developed as the leaf heats up during the drying process, and eventually the leaf temperature and moisture levels bring about the enzyme destruction or inactivation necessary for preservation of the product.

The conventional tea dryer is basically a rectangular chamber. The tea to be dried is spread on perforated trays. These dryers are driven by trays-carrying chains and sprockets running on angle iron runners. The movement of the trays with the leaf is counter-current to the draft of hot air. Recently the technology of fluid-bed drying at 125°C for 20 minutes is being followed in many cases. Normally the leaf enters the driyer with 60-72% moisture, depending on the cutting process used, and is dried down to 2.5-3.5% moisture.

Cooling , Sorting and Grading

The hot dried tea is cooled on the factory floor before it is sent for sorting and grading. Sorting is the operation in which tea particles of the bulk are separated into various grades of different sizes and forms confirming to trade requirements. In other words, it basically converts the bulk into finished products.

The process of sorting has two objectives to impart quality. Sorting enhances the appearance and quality of liquor; at the same time it can also deteriorate the quality. The presence of fibre or flakes of coarse leaf in a primary grade causes harshness and their removal makes the liquor mellow. The cleaning of fibre also improves the black appearance of tea which is desirable. Bloom is indicative of liquor character; over sorting and over cleaning can result in loss of bloom. Usually a tea which has not been well fired, loses bloom more quickly.

If tea absorbs moisture during the cleaning process, liquors can deteriorate and its keeping quality reduces. Sorting of bulk has to be done in three stages – (1) Cleaning of fibre Winnowing. The initial separation of leaf and dust can be carried out on a Middleton fitted with a sieve on No.24 mesh on the top and No.10 mesh on the bottom. Tea particles which pass through mesh No.24 are collected separately and graded further in the dust grade sorter. Similarly, particles which are retained on No. 24 mesh and pass through mesh No.10 are graded separately in the leaf grade sorter. Particles retained over 10 are generally taken for reconditioning. Both leaf and dust grades are further fractionated either in Trinic sorters or vibro sorters. After sifting, each grade is finally cleaned by winnowing during which fibres and flaky teas are removed. Generally, Super Fine Dust grades are cleaned separately in the fibrex after winnowing.

Packing

Sorted and graded black tea is normally packed in a ply wood tea chest. The chests are assembled using various components, like the plywood shooks, battens, metal fittings, iron nails and aluminium foil with tissue paper lining. Recently attention has been drawn to the use of non-conventional materials like paper, multiwall paper sacks and polythene lined jute bags.

Fuel and Power Requirements

Factories require a power source for driving machinery and a fuel source for heat generation. The nature and combination of fuel and power sources varies considerably depending on the natural resources of the tea growing area concerned, and on the economics applying to possible alternatives. The efficiency of the drying system and local meteorological conditions will also effect the fuel consumption.

Electrical Power Source

The simplest source of power is in areas where a national supply and distribution network is available. However tea estates often install their own generation facilities for running the plant and other facilities during power shutdown period. Nowadays most standby generators are driven by diesel engines, producing about 3 kWh per litre of fuel consumed. To produce one tonne of tea normally requires between 430 and 490 kWh.

TEA PLANTATION AND RELATED ASPECTS

SOIL

Tea is grown in a wide range of soil types found in tropical, subtropical and temperate climate conditions. These soil types have developed from diverse parent rock material and under different climate conditions. In China, Indonesia, Sri Lanka, South India, Turkey and Georgia (USSR) tea is mostly grown on sedimentary soils derived from gneiss or granite. In north-east India, except in Darjeeling, tea is grown on flat alluvial lands which occupy the vast area of the Brahmaputra Valley in Assam. Peat soils that have been drained are successfully used in Cachar to grow tea. But in Kyoto and Kanaya, the main areas of Japanese tea, the crop is grown on soil types derived from volcanic ash and in Taiwan, tea occupies a tract of tertiary rocks derived from a residual formation. However, all these soils have one common characteristic i.e. they grows best in acid soils. Despite the diversity of soil types on which tea is grown, all the soils exist in high rainfall conditions, as this is the most important climatic factor for successful tea growing. These soils get formed under special type of weather combined with intensive leaching of the products of weathering. The degree of leaching and hence the character of the resulting soil depends not only on the rainfall but also on the temperature. Because of differences in temperature, soils formed in tropical climates are likely to have certain characters different from those formed in the sub-tropical or semi-tropical conditions. The tea areas in tropical climates experience minimal temperature variations as compared with those in subtropical climates.

RAINFALL

The relationship between tea production and rainfall is well known. Tea grows well in areas having a precipitation of about 1,150 to 8,000 mm. The effect of rainfall is perhaps more manifested by its influence on moisture status of the soil and in inducing vegetative growth. Therefore, distribution of rainfall is as important as the total annual rainfall. It is difficult to say what exactly is the water requirement of tea in different phases of its growth and development. It is natural to expect that water requirement of tea would vary according to the prevailing environmental conditions, but it is reasonably assumed that tea on an average may transpire 900 mm per annum. Ideally, the crop water requirement should be such that it does not cause any stress in the plant system.

HUMIDITY

Humidity is of importance in tea physiology primarily because of its influence in determining the loss of moisture by evapo-transpiration. High humidity reduces water loss, but low humidity increases it, and also increases water stress during rainless drought period. In relation to temperature, low humidity may be advantageous because by increasing transpiration rate it also reduces leaf temperature, though the effect will be more pronounced in unshaded conditions. In north-east India the humidity level generally remains high during the harvesting period and this is generally considered to be conducive to growth.

AIR TEMPERATURE

Tea is grown in tropical to temperate conditions, under a regime of air temperature that varies between –8°C and 35°C; the suitable temperature for growth being the one common to the habitat of the plant, that is, where it grows. But photosynthetic rate of tea is at maximum between 30°C and 35°C, falls rapidly at 37°C, and between 39°C and 42°C there is virtually no net photosynthesis. There is also no uptake of carbon dioxide at about 42°C; respiration may continue up to about 48°C but the leaf is irrevocably damaged. Therefore, there has to be an optimum range of temperature for growth and productivity of tea. In north-east India the air temperature remains above 30°C for long periods during harvesting season, and hence the need for shade to keep the temperature below a level where it cannot affect photosynthesis adversely. It is important that a high vapour-pressure deficiency is also associated with high temperatures; and it would be useful to partition the effects of these two components on plant growth.

LAND PREPARATION

Environmental considerations discourage the replacement of rain forest by tea. However, it is worth recalling that a properly managed tea plantation causes only a temporary disturbance of many of the characteristics of the rain forest. Whilst tea cannot be rain forest as the plant species and associated fauna are different, the stream flows and evapo-transpiration revert to the same as the forest when the tea is mature. Clearance of any type of vegetation disturbs the soil and leaves it unprotected. Clearance and planting must be timed with the minimum of intervening delay. The mechanics of planting depend on the density of vegetation to be removed. Trees should be killed in advance of clearing to minimise the risk of root disease remaining alive to infect the tea. Trees should be ring barked or frilled; the trees need inspection at intervals because the bark tends to regrow across such cuts. Trees may take as long as three years to die. Chemicals used at one time to kill trees have been withdrawn for safety reasons. Trees and other heavy material should be moved off the field. The accumulation of quantities of organic material in heaps must be avoided as it may raise the pH of the soil to an unacceptable level. Light material may be ploughed in, or burnt if well spread out. Large fires must be avoided as the ash will raise the pH of the soil. A low pH will be more tolerant of ash and vegetation than soil with pH close to the upper limit for tea.

EROSION CONTROL AND DRAINS

Drains should be put in as needed on flat land. These are particularly necessary on swampy ground and should be deep enough to allow for the settlement which often occurs when such land is drained. Roadway formations should be leveled with drains and culverts put in place.

WATER REQUIREMENT

Theoretically, the water requirement of tea is the amount needed to balance the water lost by transpiration, which is a continuous process. Though rainfall is essential for growth & productivity of tea, it must be of the right quantum and intensity.

The effect of rainfall is perhaps more manifested by its influence on moisture status of the soil. Normally a part of rainwater is lost by surface run off, percolation and evaporation, only a portion of the heavy & high intensity rain percolates to the root zone of tea to cause appreciate physiological effects. Soil conditions also react with rainfall in more than one way. The ability of the soil to withstand long dry season without adverse effect depends on the quantity of rain water that the soil is capable of retaining and can make available to the plants.

This depends, apart from the capacity of the soil to retain water, on the spread and depth of tea roots. On the other hand, excessive rainfall in shallow soil and depressions may cause localised waterlogging - the latter affects both root efficiency and photosynthesis. Heavy rainfall may also affect pollination. However, the adverse effects of either excess rainfall or lack of adequate rainfall can be compensated by drainage and irrigation respectively. Normally in tea plantation water is supplied in the form of spray by water sprinkler. This measure controls the soil moisture and also allows coverage of large sectors from limited resources.

Drainage and irrigation are very closely linked with each other. A well-drained land will provide better conditions for tea roots to grow deeper and to proliferate well. A deep rooting system will thus be able to explore a much bigger volume of soil for water and nutrients and will be able to withstand drought better and longer. On the other hand, in poorly drained lands the roots remain very shallow and tea starts showing symptoms of soil moisture stress much earlier. This also demands for more frequent irrigation for survival. To obtain optimum benefits of irrigation, it is therefore essential to provide adequate drainage in the area to be irrigated.

SOIL CHEMISTRY

Acidity

It is generally agreed that optimum range of soil pH for the tea plant is 5.0 to 5.6. Soil those above pH 5.6 require correcting with acid materials such as Aluminium sulphate, Sulphate of ammonia, elemental Sulphur. On the lower end of the scale tea grows in as low a pH as 4.0 or below. The main causes of soil acidity are Hydrogen and Aluminium ions. In an intensively leached soil in which tea grows best, cations are replaced by H+ from the water molecule and can be washed off the mineral. This is the main source of acidity in tea soils. Aluminium ions also react with water, produce acidity. Nitrogenous fertilizers and decomposed organic matter in soils also supply Hydrogen ions cause acidity to soil. In case of soil pH less than 5.0, use of lime to bring the pH to about 5.0 is quite convenient.

Organic Matter

Different Organic Matters accumulate into the soil from plant & animal tissue through well decomposition, acts as the binding material joining soil particles to form a stable soil structure. Under natural condition it is the main source of plant nutrient. Tea soils vary in their OM contents ranging from less than 1% to over 30%. However, the tea plant once established and well maintained generates a lot of OM from leaf fall and prunings (10 - 20 tonnes biomas produced per hectare per year).

WATER LOGGING

Water logging is a physical factor, which can render a soil to be unsuitable for tea. This is very important because it is known that tea roots will not grow when the soil or subsoil is water logged for a prolonged period. Thus, a soil which has poor drainage should be avoided.

Water Holding Capacity

The water-holding capacity of a given tea soil and its availability to plants is dependent on soil textural composition, their aggregation and the rooting depth of the tea plant. The coarser the texture (higher content of coarse material) the less the water-holding capacity. The less the water-holding capacity the shorter times such a soil will be able to supply water to the tea plant, especially in an extended period without rain or irrigation.

FERTILIZER

Fertilisation is an important part of the normal intensive production of tea. Tea is normally grown as a long-term monoculture. Without applied fertilizer the supply of nutrients available in the soil will become exhausted leading to mineral deficiencies in the plants, severe reduction in yield and ultimately, to the death of plants and a degraded plantation. Nutrients are removed from a field of tea in a number of ways. There is an irreversible loss in the crop. The net loss of nutrients in old leaves and prunings depend on the extent to which these items are retained in the field. Soil erosion, drainage of excess water containing nutrients in solution and decomposition to gases create further losses. Uncontrolled weeds also absorb substantial quantities of nutrients. In addition to oxygen, hydrogen and carbon which are obtainable in air & water, tea plant requires fourteen (14) mineral nutrients for its ideal growth. These are Nitrogen , Potassium , Magnesium , Manganese , Copper (Cu), Aluminium , Boron and Chlorine (Cl). Of these essential nutrients mentioned above, the first nine are considered principal nutrients as they are required in relatively larger amounts. The details of plant nutrients is being presented below in Fig.

Functions Of Crop Nutrients

Primary nutrients :

- Nitrogen (N)
  - Induces vegetative growth
  - Essential constituent of protein, chlorophyll, hormones, alkaloids and Vitamins
  - Governs the ultisation of Phosphorus and Potassium

PLANT NUTRIENTS

MAJOR NUTRIENTS		MINOR NUTRIENTS	
Primary Nutrients	**Secondary Nutrients**	**Micro Nutrients**	**Beneficial elements**
Carbon	Magnesium	Zinc	Aluminium
Hydrogen	Calcium	Manganese	Silicon
Oxygen	Sulphur	Boron	
Nitrogen		Iorn	
Phosphorus		Copper	
Potassium		Molybdenum	
		Chlorine	

- Phosphorus (P) :
  - Stimulates root growth
  - Carrier of energy
  - Essential for synthesis of starch, protein and fat

  - Constituent of nucleic acids and enzymes
  - Related to cell division and development
- Potassium (K) :
  - Essential for frame development
  - Regulates water metabolism in plants
  - Provides resistance against drought, frost, pests and diseases
  - Acts as an accelerator of enzyme action
  - Counteracts the ill-effects of excess Nitrogen

Secondary nutrients :

- Calcium (Ca) :
  - Constituent of cell wall
  - Necessary for normal mitosis
  - Essential to active growing points
  - Activator of enzymes
- Magnesium (Mg) :
  - Constituent of chlorophyll
  - Imparts dark green colour to leaves
  - Helps in synthesis of carbohydrates, protein, fat and vitamins
- Sulphur (S) :
  - Constituent of Sulphur bearing amino acids
  - Helps in stabilizing protein structure

Deficiency Symptoms Of Nutrients

Primary nutrients :

- Nitrogen (N) :
  - Young flush shows a lighter than normal green colour-yellowing
  - Marked decline in crop
- Phosphorus (P) :
  - Mature leaves loose their glossiness and appear dull
  - Stunted appearance of plants
- Potassium (K) :
  - Marginal necrosis in mature leaves
  - Premature leaf fall leaving a crown of tender foliage
  - Development of thin and twiggy wood
  - Poor recovery after plucking and pruning

Secondary nutrients:

- Magnesium :
  - Seen on mature leaves
  - Leaves appear bright yellow with a conspicuous inverted dark green "V" down the
  - Mid rib
- Sulphur :
  - Chlorotic symptoms occur in younger and recently formed leaves

- Tea
- Yellows
- Lateral extension of leaves is restricted leading to formation of narrow laves

- Micronutrients :
- Boron :
  - Terminal bud remains dormant and die early
  - Clusters of small shoots develop in the upper axils
  - Development of translucent oil spots on the lower surface of mature leaves
  - Formation of cork mound on the leaf stalk

Organic Fertilizer

Organic manure is applied to tea as green manure or as one of many other organic materials. Organic materials limits the deterioration of the soil and improves soil fertility and structure. A bed of humus is formed which supports an active micro fauna which speeds decomposition converting nutrients to an available from. Erosion is reduced and the infiltration of water improved.

Green Manure

The tea plants provide some green manure from leaf-fall. The leaves of shade also form green manure. Leguminous or other trees planted in lines as windbreaks are trimmed frequently in some areas; leguminous trees, planted as hedges separated by a few lines of young tea as temporary shelter, can provide up to 100 kg Nitrogen per hectare per annum within 25-30 tonnes organic matter. These close hedges have to be removed before the tea comes into production. Removal of the prunings takes a large quantity of nutrients out of the plant soil system. This loss will not be replaced by normal fertilizer applications. If prunings are removed regularly the tea will develop nutrient deficiencies, yield will fall and the health of the plants will decline. Grice and Malenga reported that removal of prunings produced a 4% reduction in crop yield in the cycle after a down prune.

Bibliography

A C Mittal: *Agro-Rural Sector and Microfinance*, Arise Publication, Delhi, 2008.

Avtar Singh Bimbraw: *Agro-techniques for Umbelliferous Medicinal and Aromatic Plants of India*, International Book Distributing Co, 2006.

B.S. Yadav and Sachi Rana: *Rural Empowerment Through Agro-Based Industries*, Shree Publication, Delhil, 2005.

D. Das Gupta and S. Chowdhury: *Agro-Technology Transfer Programme—An Assessment*, Mittal, Publication, Delhi, 2002.

Deepa H. Dwivedi, S.K. Dwivedi and S. Gupta: *Agro-Animal Resources of Higher Himalayas*, Satish Serial Publication, Delhi, 2010.

Deepa H. Dwivedi, S.K. Dwivedi: *Agro-Animals Resources of Higher Himalayas*, S. Gupta, Satish Serial Publication, House, 2010.

G. Vanitha and M. Kalpana: *Agro-Informatics*, New India Publishing Agency, 2011.

G.V. Joshi: *Agro-Based Industries: Problems and Prospects* , Mohit, Publication, Delhi, 2002.

Gouradevi S. Katnalli: *Agro-Based Industries in India: An Economic Analysis*, ABD Publication, Delhi, 2011.

H C Srivastava: *Valuable Directory to Boost Medicinal and Aromatic Agro-Industries in India*, International Book Dist, 2008.

H C Srivastava: *Valuable Directory to Boost Medicinal and Aromatic Agro-Industries in India*, International Book Dist, 2008.

H.P Singh and M.S. Palaniswami: *Horticulture in Different Agro-Climatic Conditions : Four Decades of Coordinated Research,* Westville Publishing House, 2010.

H.R. Yadav: *Agro-Afforestation Management on Wastelands : Village Level Study*, Concept, Publication, Delhi, 2011.

Keshav Dev Gaur and Anil Kumar Singh: *Economics of Agro-Industries*, Sunrise, Publication, Delhi, 2008.

Keshav Dev Gaur and Anil Kumar Singh: *Economics of Agro-Industries* , Sunrise, Publication, Delhi, 2008.

Mun Mun Roy: *Changing Patterns of Sex Ratio of the Agro-Climatic Zones of India*, Sampark, Publication, Delhi, 2010.

Nandini Sinha Kapur: *Reconstructing Identities : Tribes, Agro-Pastoralists and Environment in Western India (Seventh-Twentieth Centuries)*, Manohar, Publication, Delhi, 2008.

P K Behera: *Plant Compost-Manure and Agro-Chemicals Analysis : A Laboratory Manual (2 Vols-Set)* , Dominant, Publication, Delhi, 2007.

R. Sasikumar: *Food Processing Technology in Agro-Based Sector*, Biotech Books, 2014.

Robert D Fitzgerald and Charles A Gardner: *Encyclopaedia of Organic Agro-Forestry, Vols. 1-3* Dominant Publication, Delhi, 2009.

S N Misra: *Rural Development Through Agro-Industries*, Anmol, Publication, Delhi, 2004.

S N Misra: *Rural Development Through Agro-Industries*, Anmol, Publication, Delhi, 2004.

S. Solomon, S.S. Grewal, Yang-rui Li, R.C. Magarey and G.P. Rao: *Sugar Cane : Production Management and Agro-Industrial Imperatives*, International, 2005.

Sadhana Singh: *Agro-Techniques of Medicinal Plants*, Wisdom Press, 2012.

Y. Sreenivasulu: *Functioning of Land Lease Market in Different Agro-Climatic Settings in India*, Serials Publications, 2015.

Index